GALAXIES MASSES AS CONSTRAINTS OF FORMATION MODELS

IAU SYMPOSIUM No. 311

COVER ILLUSTRATION: CONFERENCE POSTER

The cover picture is the official conference poster for the IAU Symposium 311. It shows the face-on spiral galaxy NGC 1232 looking down on the colleges in Oxford UK, where the conference was held.

This choice of a spiral galaxy reminds us of the first targets for which galaxy masses were first measured in the sixties from major axis rotation curves. This matches our conference theme which addresses the interplay between galaxy masses and galaxy evolution.

The spiral galaxy image was taken at ESO. The Oxford skyline is by former Oxford student Joseph Caruana and is reproduced with permission. The poster was designed by Oxford graduate student Shravan Shetty.

IAU SYMPOSIUM PROCEEDINGS SERIES

INTERNATIONAL ASTRONOMICAL UNION
UNION ASTRONOMIQUE INTERNATIONALE

International Astronomical Union

GALAXIES MASSES AS CONSTRAINTS OF FORMATION MODELS

PROCEEDINGS OF THE 311th SYMPOSIUM OF THE INTERNATIONAL ASTRONOMICAL UNION HELD IN OXFORD, UNITED KINGDOM JULY 21–25, 2014

Edited by

MICHELE CAPPELLARI
Sub-department of Astrophysics, Department of Physics, University of Oxford, Denys Wilkinson Building, Keble Road, Oxford OX1 3RH, UK

and

STÉPHANE COURTEAU
Queen's University, Department of Physics, Engineering Physics and Astronomy, Kingston, Ontario, Canada K7L 3N6

CAMBRIDGE
UNIVERSITY PRESS

CAMBRIDGE UNIVERSITY PRESS
University Printing House, Cambridge CB2 2BS, United Kingdom
40 West 20th Street, New York, NY 10011–4211, USA
10 Stamford Road, Oakleigh, Melbourne 3166, Australia

First published 2015

Printed in the UK by Bell & Bain, Glasgow, UK

Typeset in System LaTeX 2ε

A catalogue record for this book is available from the British Library Library of Congress Cataloguing in Publication data

ISBN 9781107078697 hardback
ISSN 1743-9213

Table of Contents

Preface

In the era of precision cosmology, we think we can accurately predict the distribution of dark matter in the Universe. However the impact of baryonic physics is still largely unknown and our understanding of galaxy formation must rely on observations. A key advance in recent years has been the ability to enrich studies of the luminosity evolution of galaxies with determinations of their stellar or total masses from dynamical analyses using stellar populations, stellar or gaseous dynamical models, weak or strong lensing. Contrary to the light distribution alone, the distribution of both the stellar and dark matter in galaxies can be robustly compared to galaxy formation models.

Dynamical studies of galaxies near and far have evolved from modelling the mass distribution of individual objects to capitalizing on large surveys using integral field and multi-object spectroscopy, strong or weak gravitational lensing, planetary nebulae, stellar and gas kinematics, and multi-wavelength studies, to constrain masses from the stellar population. Much of this progress has relied on key instrumentation developments. For instance, new spectrographs optimized to near-infrared wavelengths now better trace the rest-frame visual spectra of distant galaxies. Massive multi-objects capabilities also allow larger samples to be obtained in feasible exposure times. In the foreseeable future, 30–40-m class telescopes, the LSST survey and JWST and EUCLID missions promise to extend our studies of galaxy masses and kinematics of nearby galaxies up to redshift $z \sim 2$ and beyond, where most of the galaxy assembly has taken place.

This symposium aims at bringing together galaxy evolution theorists, observers of the nearby and distant universe, and instrumentation specialists. We must identify what key observables can be robustly reproduced by the models, how the existing and new instrumentation should be optimized for galaxy evolution studies, and what future observations would be most useful to constrain the models.

This Symposium was also an opportunity to celebrate the illustrious career of Prof. Roger Davies.

Michele Cappellari and Stéphane Courteau, co-chairs SOC
Oxford and Kingston, 1 January, 2015

THE ORGANIZING COMMITTEE

Scientific

Roland Bacon (France)
Joss Bland-Hawthorn (Australia)
Alyson Brooks (USA)
Kevin Bundy (Japan)
Michele Cappellari (co-chair, UK)
Charlie Conroy (USA)
Stéphane Courteau (co-chair, Canada)
Gabriella De Lucia (Italy)
Natascha Förster Schreiber (Germany)
Claudia Maraston (UK)
Amélie Saintonge (UK)
Alice Shapley (USA)
Tommaso Treu (USA)
Ignacio Trujillo (Spain)
Frank van den Bosch (USA)
Sukyoung Yi (Korea)

Local

Martin Bureau
Michele Cappellari (chair)
Ryan Houghton
Sugata Kaviraj
Leanne O'Donnell
John Magorrian
Marc Sarzi
Shravan Shetty
Alfred Tiley

Acknowledgements

The symposium is sponsored and supported by the IAU Divisions B (Facilities, Technologies and Data Science), G (Stars and Stellar Physics), H (Interstellar Matter and Local Universe) and J (Galaxies and Cosmology).

The Local Organizing Committee operated under the auspices of the Sub-dept. of Astrophysics, Dept. of Physics of the University of Oxford

Funding by the
International Astronomical Union
and by the University of Oxford,
is gratefully acknowledged.

CONFERENCE PHOTOGRAPH

Participants

Roberto **Abraham**, University of Toronto, CANADA	abraham@astro.utoronto.ca
Louis **Abramson**, University of Chicago, USA	labramson@uchicago.edu
Padrai **Alton**, Durham University, UK	padraig.alton@durham.ac.uk
Josh **Argyle**, University of St. Andrews, UK	ja66@st-andrews.ac.uk
Nobuo **Arimoto**, Subaru Telescope, NAOJ, Hawaii	arimoto@naoj.org
Madga **Arnaboldi**, European Southern Observatory, GERMANY	marnabol@eso.org
Roland **Bacon**, CRAL, FRANCE	rmb@obs.univ-lyon1.fr
Peter **Behroozi**, Space Telescope Science Institute, USA	pbehroozi@gmail.com
Alessandra **Beifiori**, Max Plank Institute for Astrophysics, GERMANY	beifiori@mpe.mpg.de
Sirio **Belli**, Caltech, USA	sirio@astro.caltech.edu
Rachel **Bezanson**, University of Arizona/Steward Observatory, USA	rbezanson@email.arizona.edu
James **Binney**, University of Oxford, UK	binney@thphys.ox.ac.uk
Joss **Bland Hawthorn**, University of Sydney, AUSTRALIA	jbh@physics.usyd.edu.au
Asa **Bluck**, University of Victoria, CANADA	abluck@uvic.ca
Nicholas **Boardman**, University of St. Andrews, UK	nfb@st-andrews.ac.uk
Ana **Bonaca**, Yale University, USA	ana.bonaca@yale.edu
Jo **Bovy**, Institute for Advanced Study, USA	bovy@ias.edu
Jean **Brodie**, University of California - Santa Cruz, USA	jbrodie@ucsc.edu
Margot **Brouwer**, Leiden Observatory, The Netherlands	margot.brouwer@gmail.com
Stewart **Buchan**, University of Southampton, UK	s.w.buchan@soton.ac.uk
James **Bullock**, UC Irvine, USA	bullock@uci.edu
Kevin **Bundy**, Kavli IPMU, JAPAN	kevin.bundy@ipmu.jp
David **Buote**, University of California - Irvine, USA	buote@uci.edu
Martin **Bureau**, University of Oxford, UK	bureau@astro.ox.ac.uk
Marcello **Cacciato**, Leiden Observatory, The Netherlands	cacciato@strw.leidenuniv.nl
Michele **Cappellari**, University of Oxford, UK	cappellari@astro.ox.ac.uk
Maria **Cebrian Renau**, Instituto de Astrofísica de Canarias, SPAIN	mcebrian@iac.es
Gilles **Chabrier**, ENS-Lyon, FRANCE	chabrier@ens-lyon.fr
Aldée **Charbonnier**, Universidade Federal do Rio de Janeiro (UFRJ), BRAZIL	charbonnier@astro.ufrj.br
Stéphane **Charlot**, Institut d'Astrophysique de Paris, FRANCE	charlot@iap.fr
Yanping **Chen**, New York University Abu Dhabi (NYUAD), UNITED ARAB EMIRATES	chenyp.astro@gmail.com
Charlie **Conroy**, University of California - Santa Cruz, USA	conroy@ucsc.edu
Guido **Consolandi**, University of Milano Bicocca, ITALY	guido.consolandi@mib.infn.it
Arianna **Cortesi**, IAG-USP, BRAZIL	aricorte@gmail.com
Stéphane **Courteau**, Queen's University, CANADA	courteau@astro.queensu.ca
Scott **Croom**, University of Sydney, AUSTRALIA	scroom@physics.usyd.edu.au
Emanuele **Daddi**, CEA Saclay, FRANCE	edaddi@cea.fr
Iary **Davidzon**, Università di Bologna, ITALY	iary.davidzon@unibo.it
Roger **Davies**, University of Oxford, UK	rld@astro.ox.ac.uk
Timothy **Davis**, European Southern Obsdervatory, GERMANY	tdavis@eso.org
Roelof **de Jong**, Leibniz-Institut für Astrophysik Potsdam (AIP) , GERMANY	rdejong@aip.de
Adriana **de Lorenzo-Cáceres**, University of St. Andrews, UK	adlcr@st-andrews.ac.uk
Avishai **Dekel**, The Hebrew University of Jerusalem, ISRAEL	ekel@huji.ac.il
Denedikt **Diemer**, University of Chicago / KICP, USA	bdiemer@oddjob.uchicago.edu
Alan **Dressler**, Carnegie Observatories, USA	dressler@obs.carnegiescience.edu
Richard **D'Souza**, Max Plank Institute for Astrophysics, GERMANY	rdsouza@mpa-garching.mpg.de
Pierre-Alain **Duc**, AIM Paris - Saclay, FRANCE	paduc@cea.fr
Aaron **Dutton**, Max Planck Institute for Astronomy, GERMANY	dutton@mpia.de
Sandra **Faber**, University of California Observatories, USA	faber@ucolick.org
Jesus **Falcon-Barroso**, Instituto de Astrofísica de Canaries, SPAIN	jfalcon@iac.es
Michael **Fall**, STScI, USA	fall@stsci.edu
Anna **Ferré-Mateu**, Subaru Telescope, USA	aferre@naoj.org
Duncan **Forbes**, Swinburne University, AUSTRALIA	dforbes@swin.edu.au
Francesca **Fragkoudi**, LAM, FRANCE	francesca.fragkoudi@lam.fr
Carols **Frenk**, ICC - Durham, UK	c.s.frenk@durham.ac.uk
Jay **Frogel**, Galaxies Unlimited and World Images, USA	jayfrogel@me.com
Dimitri **Gadotti**, European Southern Observatory, CHILE	dgadotti@eso.org
Anna **Gallazzi**, INAF-Osservatorio Astrofisico di Arcetri, ITALY	gallazzi@arcetri.astro.it
Reinhard **Genzel**, Max Planck Institute for Extraterrestrial Physics, GERMANY	genzel@mpe.mpg.de
Ortwin **Gerhard**, MPE, GERMANY	gerhard@mpe.mpg.de
Boris **Haeussler**, University of Oxford, UK	BorisHaeussler.astro@gmail.com
Peter **Hague**, University of Leicester, UK	peter.hague@le.ac.uk
Claire **Halliday**, FRANCE	claire.halliday@free.fr
Jiaxin **Han**, ICC - Durham University, UK	hanjiaxin@gmail.com
David **Hanes**, Queen's University, CANADA	hanes@astro.queensu.ca
Kohei **Hayashi**, Tohoku University, JAPAN	k.hayasi@astr.tohoku.ac.jp
Amina **Helmi**, Kapteyn Institute, The Netherlands	ahelmi@astro.rug.nl
Maren **Hempel**, Pontificia Universidad Católica de Chile, CHILE	maren551@gmail.com
Ana **Hidalgo-Gámez**, Escuela Superior de Fisica y Matematicas - IPN, MEXICO	ahidalgo@esfm.ipn.mx
Michaela **Hirschmann**, INAF-Trieste/IAP-Paris, ITALY	mhirsch@oats.inaf.it
Henk **Hoekstra**, Leiden Observatory, The Netherlands	hoekstra@strw.leidenuniv.nl
Ryan **Houghton**, University of Oxford, UK	rcwh@astro.ox.ac.uk
Mei-Ling **Huang**, Max-Planck Institute for Astrophysics, GERMANY	mlhuang@MPA-Garching.MPG.DE
Mike **Hudson**, University of Waterloo, USA	mjhudson@uwaterloo.ca
Marc **Huertas-Company**, Paris Observatory, FRANCE	marc.huertas@obspm.fr
Wako **Ishibashi**, ETH Zurich Institute for Astronomy, SWITZERLAND	wako.ishibashi@phys.ethz.ch
Prajwal **Kafle**, The University of Sydney, AUSTRALIA	p.kafle@physics.usyd.edu.au
Emin **Karabal**, ESO / Cea Saclay, FRANCE	karabalemin@gmail.com
Susan **Kassin**, Space Telescope Science Institute, USA	kassin@Stsci.edu
Guinevere **Kauffmann**, Max Planck Institute for Astrophysics, GERMANY	gamk@mpa-garching.mpg.de
Sugata **Kaviraj**, University of Hertfordshire, UK	s.kaviraj@herts.ac.uk
Sarah **Kendrew**, University of Oxford, Oxford, UK	sarah.kendrew@astro.ox.ac.uk
Chiaki **Kobayashi**, University of Hertfordshire, UK	c.kobayashi@herts.ac.uk
John **Kormendy**, University of Texas at Austin, USA	kormendy@astro.as.utexas.edu
Davor **Krajnovic**, Leibniz-Institut für Astrophysik Potsdam (AIP), GERMANY	davor@aip.de
Mariska **Kreik**, UC Berkeley, USA	mkriek@berkeley.edu

Ulrike **Kuchner**, University of Vienna, AUSTRIA	ulrike.kuchner@univie.ac.at
Harald **Kuntschner**, ESO, GERMANY	hkuntsch@eso.org
Francesco **La Barbera**, INAF-OAC, ITALY	labarber@na.astro.it
Claire **Lackner**, Kavli IPMU, JAPAN	claire.lackner@ipmu.jp
Chervin **Laporte**, Max Planck Institute for Astrophysics, GERMANY	cfpl2@mpa-garching.mpg.de
Ronald **Lasker**, MPIA, GERMANY	laesker@mpia.de
Alexie **Leauthaud**, Kavli IPMU, JAPAN	alexie.leauthaud@ipmu.jp
Federico **Lelli**, Case Western Reserve University, USA	federico.lelli@case.edu
Timothy **Licquia**, University of Pittsburgh, USA	tcl15@pitt.edu
Richard **Long**, NAOC, UK	rjl2007@gmail.com
Alessia **Longobardi**, Max Planck Institute for extraterrestrial Physics, GERMANY	alongobardi@mpe.mpg.de
John **Lucey**, Durham University, UK	john.lucey@durham.ac.uk
Donald **Lynden-Bell**, IOA - Cambridge, UK	dlb@ast.cam.ac.uk
Natalya **Lyskova**, Max Planck Institute for Astrophysics, GERMANY	lyskova@mpa-garching.mpg.de
Mariya **Lyubenova**, Kapteyn Astronomycal Institute, The Netherlands	lyubenova@astro.rug.nl
John **Magorrian**, University of Oxford, UK	magog@thphys.ox.ac.uk
Millicent **Maier**, Australian Astronomical Observatory, AUSTRALIA	mmaier@aao.gov.au
Gary **Mamon**, IAP, FRANCE	gam@iap.fr
Alison **Wing Shan Man**, Dark Cosmology Centre - Niels Bohr Institute, DENMARK	allison@dark-cosmology.dk
Rachel **Mandelbaum**, Carnegie Mellon University, USA	rmandelb@andrew.cmu.edu
Nir **Mandelker**, Hebrew University of Jerusalem, ISRAEL	nir.mandelker@mail.huji.ac.il
Claudia **Marston**, ICG-University of Portsmouth, UK	claudia.maraston@port.ac.uk
Danilo **Marchesini**, Tufts University, USA	danilo.marchesini@tufts.edu
Ignacio **Martan Navarro**, Ignacio Instituto de Astrofísica de Canarias, SPAIN	imartin@iac.es
Charlotte **Mason**, University of California - Santa Barbara, USA	cmason@physics.ucsb.edu
Richard **McDermid**, Macquarie University / AAO, AUSTRALIA	richard.mcdermid@mq.edu.au
Robert **McMahon**, Kettering University, USA	mcmahan@kettering.edu
Jairo **Mendez Abreu**, University of St. Andrews, UK	jma20@st-andrews.ac.uk
Sarah **Miller**, University of California - Irvine, USA	shmiller@uci.edu
Carl **Mitchell**, Rutgers - The State University of New Jersey, USA	cmitchell@physics.rutgers.edu
Peter **Mitchell**, Durham University, UK	peter.mitchell@durham.ac.uk
Takahiro **Morshita**, Astronomical Institute of Tohoku University, JAPAN	mtakahiro@astr.tohoku.ac.jp
Laura **Morselli**, Excellence Cluster, GERMANY	laura.morselli@tum.de
Matt **Mountain**, Space Telescope Science Institute, USA	mmountain@stsci.edu
Adam **Muzzin**, Leiden Observatory, The Netherlands	muzzin@strw.leidenuniv.nl
Julio **Navarro**, Univesity of Victoria, CANADA	jfn@uvic.ca
Andrea **Negri**, Università di Bologna , ITALY	andrea.negri@unibo.it
Alberto **Nigoche-Netro**, Universidad de Guadalajara, MEXICO	anigoche@gmail.com
Carlo **Nipoti**, Università di Bologna, ITALY	carlo.nipoti@unibo.it
Go **Ogiya**, University of Tsukuba, GERMANY	ogiya@ccs.tsukuba.ac.jp
Kyoko **Onishi**, The Graduate University for Advanced Studies/NAOJ, JAPAN	kyoko.onishi@nao.ac.jp
Nathalie **Ouellette**, Queen's University, CANADA	nouellette@astro.queensu.ca
Milena **Pawlik**, University of St. Andrews, UK	mp84@st-andrews.ac.uk
Reynier **Peletier**, Kapteyn Astronomical Institute, The Netherlands	peletier@astro.rug.nl
Arriba Luis de **Peralta**, Instituto de Astrofísica de Canarias, SPAIN	lperalta@iac.es
Pablo **Pérez-González**, Universidad Complutense de Madrid, SPAIN	pgperez@ucm.es
Gabriele **Pezzulli**, Università di Bologna, ITALY	gabriele.pezzulli@unibo.it
Silvia **Posacki**, Università di Bologna, ITALY	silvia.posacki@unibo.it
Lorenzo **Posti**, Università di Bologna, ITALY	lorenzo.posti@unibo.it
Rhea-Silvia **Remus**, University Observatory Munich, GERMANY	rhea@usm.lmu.de
Alvio **Renzini**, INAF-Osservatorio Astronomico di Padova, ITALY	alvio.renzini@oapd.inaf.it
Joel **Roediger**, University of California - Santa Cruz, USA	joel.roediger@nrc-cnrc.gc.ca
Aaron **Romanowsky**, San Jose State University, USA	aaron.romanowsky@sjsu.edu
Martin **Sahlan**, University of Oxford, SWEDEN	martin.sahlen@astro.ox.ac.uk
Amélie **Saintonge**, UCL, UK	a.saintonge@ucl.ac.uk
Marc **Sarzi**, University of Hertfordshire, UK	m.sarzi@herts.ac.uk
Till **Sawala**, Institute for Computational Cosmology - Durham, UK	till.sawala@durham.ac.uk
Matthieu **Challer**, Institute for Computational Cosmology - Durham, UK	matthieu.schaller@durham.ac.uk
Paul **Schechter**, MIT Kavali Institute, USA	schech@mit.edu
Ricardo **Schiavon**, Astrophysics Research Institute - LJMU, UK	r.p.schiavon@ljmu.ac.uk
Nicholas **Scott**, The University of Sydney, AUSTRALIA	nscott@physics.usyd.edu.au
Marja **Seidel**, Instituto de Astrofísica de Canaries, SPAIN	mseidel@iac.es
Francesco **Shankar**, University of Southampton, UK	F.Shankar@soton.ac.uk
Shravan **Shetty**, University of Oxford, UK	Shravan.Shetty@astro.ox.ac.uk
Jonathan **Sick**, Queen's University, CANADA	jsick@astro.queensu.ca
Raymond **Simons**, Johns Hopkins University, USA	rsimons@pha.jhu.edu
Russell **Smith**, University of Durham, UK	russell.smith@durham.ac.uk
Chiara **Spinello**, Max-Planck-Institut für Astrophysik (MPA), GERMANY	spini@mpa-garching.mpg.de
Chuck **Steidel**, Caltech, USA	ccs@astro.caltech.edu
Massimo **Stiavelli**, Space Telescope Science Institute, USA	mstiavel@stsci.edu
Trinidad **Tapia**, Universidad Nacional Autónoma de México, MEXICO	ttapia@astro.unam.mx
James **Taylor**, University of Waterloo, CANADA	taylor@uwaterloo.ca
Matthew **Taylor**, Pontificia Universidad Católica de Chile, CHILE	mtaylor@astro.puc.cl
Philip **Taylor**, University of Hertfordshire, UK	p.taylor7@herts.ac.uk
Niranjan **Thatte**, University of Oxford, UK	thatte@astro.ox.ac.uk
Daniel **Thomas**, University of Portsmouth, UK	daniel.thomas@port.ac.uk
Jens **Thomas**, Max-Planck Institute for Extraterrestrial Physics, GERMANY	jthomas@mpe.mpg.de
Alfred **Tiley**, University of Oxford, UK	Alfred.Tiley@astro.ox.ac.uk
Tommaso **Treu**, University of California - Los Angeles, USA	tt@astro.ucla.edu
Ignacio **Trujillo**, Instituto de Astrofísica de Canarias, SPAIN	trujillo@iac.es
Athanasia **Tsati**, MPIA, GERMANY	tsatsi@mpia.de
Glenn **van de Ven**, MPIA, GERMANY	glenn@mpia.de
Frank **van den Bosch**, Yale University, USA	frank.vandenbosch@yale.edu
Remco **van den Bosch**, MPIA, GERMANY	bosch@mpia.de
Roeland **van der Marel**, STScI, USA	marel@stsci.edu
Pieter **van Dokkum**, Yale University, USA	pieter.vandokkum@yale.edu
Edo **van Uitert**, Argelander Institute for Astronomy, GERMANY	vuitert@astro.uni-bonn.de
Alexandre **Vazdekis**, Instituto de Astrofísica de Canarias, SPAIN	vazdekis@iac.es
Marc **Verheijen**, Kapteyn Astronomical Institute, The Netherlands	verheyen@astro.rug.nl
Aprajita **Verma**, University of Oxford, UK	averma@astro.ox.ac.uk

Massimo **Viola**, Leiden Observatory, The Netherlands viola@strw.leidenuniv.nl
Karina **Voggel**, ESO, GERMANY kvoggel@eso.org
Benedetta **Vulcani**, Kavli IPMU, JAPAN benedetta.vulcani@ipmu.jp
David **Wake**, The Open University, UK david.wake@open.ac.uk
Wenting **Wang**, Institute for Computational Cosmology, UK bilinxing.wenting@gmail.com
Anne-Marie **Weijmans**, University of St. Andrews, UK amw23@st-andrews.ac.uk
Kyle **Wesfall**, Kapteyn Astronomical Institute, The Netherlands westfall@astro.rug.nl
Simon **White**, Max-Planck-Institut für Astrophysik, GERMANY swhite@mpa-garching.mpg.de
Stephen **Wilkins**, Astronomy Centre - University of Sussex, UK s.wilkins@sussex.ac.uk
Sukyoung **Yi**, Yonsei University, KOREA yi@yonsei.ac.kr
Akin **Yildirim**, Max Planck Institute for Astronomy, GERMANY yildirim@mpia.de
Ling **Zhu**, MPIA, GERMANY lzhu@mpia-hd.mpg.de
Stefano **Zibetti**, INAF-Osservatorio Astrofisico di Arcetri, ITALY zibetti@arcetri.astro.it
Bodo **Ziegler**, University of Vienna, AUSTRIA bodo.ziegler@univie.ac.at
Simon **Zieleniewski**, University of Oxford, UK simon.zieleniewski@physics.ox.ac.uk

CONFERENCE PICTURES

Figure 1. LOC mavericks: Leanne O'Donnell (front) and Ashling Morris (back)

Figure 2. Department of Physics Martin Wood Lecture Theatre

Figure 3. Pieter van Dokkum and Mariska Kriek

Figure 4. After hours celebration dinner

Figure 5. Four of the Seven Samurai: Sandra Faber, Roger Davies, Alan Dressler, Donald Lynden-Bell

Figure 6. Roger Davies, Michele Cappellari, Roelof de Jong, Stéphane Courteau, Martin Bureau, Bodo Ziegler

Figure 7. Symposium banquet at Wadham College

Figure 8. Winners for best poster/short talk with symposium co-chairs. Missing: third winner Kyle Westfall. Stéphane Courteau, Marja Seidel, Athanasia Tsatsi, Michele Cappellari.

Figure 9. Symposium photographer Dr. Jay Frogel (World Images). Photo: Alvio Renzini

Figure 10. Roger and Yo Davies with a cake replica of the KPNO Mayall telescope. Millicent Maier is thanked for an inspiring cake selection. Photo: Stéphane Courteau

Galaxy Masses as Constraints of Formation Models
Proceedings IAU Symposium No. 311, 2014
M. Cappellari & S. Courteau, eds.

doi:10.1017/S1743921315003294

Local Group Proper Motion Dynamics

Roeland P. van der Marel

Space Telescope Science Institute, 3700 San Martin Drive, Baltimore, MD 21218

Abstract. Our knowledge of the dynamics and masses of galaxies in the Local Group has long been limited by the fact that only line-of-sight velocities were observationally accessible. This introduces significant degeneracies in dynamical models, which can only be resolved by measuring also the velocity components perpendicular to the line of sight. However, beyond the solar neighborhood, the corresponding proper motions have generally been too small to measure. This has changed dramatically over the past decade, especially due to the angular resolution and stability available on the Hubble Space Telescope. Proper motions can now be reliably measured throughout the Local Group, as illustrated by, e.g., the work of the HSTPROMO collaboration. In this review, I summarize the importance of proper motions for Local Group science, and I describe the current and future observational approaches and facilities available to measure proper motions. I highlight recent results on various Milky Way populations (globular clusters, the bulge, the metal-poor halo, hypervelocity stars, and tidal streams), dwarf satellite galaxies, the Magellanic Clouds and the Andromeda System.

1. Introduction: Local Group Dynamics and the Importance of Proper Motions

Structures in the Universe cluster on various scales. Our Milky Way Galaxy (MW) belongs to a small group, called the Local Group (LG). The two dominant galaxies in the LG are spiral galaxies: the MW and the Andromeda Galaxy (M31). Each of these galaxies has a significant companion that is roughly one-tenth of its mass: the MW has the Magellanic Clouds, i.e. the pair of the Large and Small Magellanic Clouds (LMC and SMC), and M31 has the Triangulum Galaxy (M33). The MW and M31 also each have their own system of dwarf satellite galaxies, which have much lower mass.

The Universe evolves hierarchically, with small structures merging and falling in to form bigger structures. Owing to its proximity, the LG is the best place to study some of these processes in detail. The last decade has seen a wealth of new discoveries in this area. For example, the ongoing disruption of the Sagittarius dwarf spheroidal (Sgr dSph) galaxy has produced a giant stream of stars around the Milky Way. The SDSS survey has revealed many other such streams, and there is also a giant stellar stream around M31. For these reasons, the LG and its two dominant spirals have become the benchmark for testing many aspects of galaxy formation and cosmological theories.

To understand the LG, it is necessary to study the dynamics of its stars and galaxies. The dynamics contain an imprint of the initial conditions, and also reflect subsequent evolution. Moreover, dynamical measurements are necessary to constrain the amount of (dark) mass, since the dynamics, structure, and mass of galaxies are tied through the laws of gravity and dynamical equilibrium.

Almost everything that is known about the dynamics of the LG, and of galaxies in the Universe in general, is based on observations of line-of-sight (LOS) velocities. Such observations constrain only one component of motion, and interpretation therefore generally requires that various assumptions be made. This limits the amount of insight that can be gained, e.g., through the well-known degeneracy between mass and velocity dispersion anisotropy (Binney & Mamon 1987).

An important step forward is to determine fully three-dimensional velocities, by also measuring proper motions (PMs) in the plane of the sky. This is possible with several techniques, as discussed in Section 2. PM measurements over the past decade have started to revolutionize our understanding of the LG, as reviewed (roughly in order of increasing distance) in Sections 3–6. Concluding remarks are presented in Section 7.

2. Observational Approaches and Future Prospects

High PM accuracy is required to meaningfully address most dynamical topics in the LG. For reference, Δ PM $= 50\,\mu$as/yr corresponds to a velocity accuracy $\Delta v \approx (D/4)$ km/s at distance D kpc. Therefore, this is the accuracy required to study, e.g., MW halo objects at 40 kpc with 10 km s^{-1} accuracy, or to study dwarf satellite galaxies at 200 kpc with 50 km s^{-1} accuracy. To properly grasp the challenge that this poses, it is worth noting that $50\,\mu$as/yr corresponds roughly to the angular velocity with which one would observe the hair grow of a human placed at the distance of the moon.

The highest PM accuracies, Δ PM $\lesssim 10\,\mu$as/yr, are generally obtained using interferometric telescope arrays at radio wavelengths. However, the number of sources in the LG that can be studied in this way is small. Water masers are the main targets, and these exist only in the few LG galaxies with high star formation rates. PM measurements have been reported for M33 and IC10, both satellites of M31 (Brunthaler *et al.* 2005, 2007). In the future, such measurements may become possible also for, e.g., M31 (Darling 2011) and the Magellanic Clouds.

Ground-based optical PM measurements have traditionally been based on comparison of photographic plate data taken at widely separated epochs (decades apart; e.g., Carlin *et al.* 2012). A variant of this is to compare historical photographic plate data with recent targeted CCD data (e.g., Casetti-Dinescu *et al.* 2007). Another approach is to compare homogeneous data from large surveys (e.g., SDSS) using shorter time baselines (e.g., Koposov *et al.* 2013). The accuracies thus reached are comparable to what is obtained from near-IR measurements for stars near the Galactic Center (e.g., Ghez *et al.* 2008). However, those stars move unusually fast around the MW's supermassive black hole, and they are relatively nearby. The ground-based techniques do not generally reach PM accuracies *on a per-star basis* that are sufficient for studies of Local Group dynamics. However, by calculating the average PM of large numbers of N stars, it is possible to reach the required accuracies (since random errors scale as $N^{-0.5}$).

The highest-quality optical/near-IR PM measurements are obtained from space, and specifically with the Hubble Space Telescope (HST). HST's location above the earth atmosphere yields several advantages, including high spatial resolution, low sky background, and exquisite telescope stability. Moreover, there is a large data archive of observations spanning 24 years (or 12 years when restricted to the highest quality cameras ACS and WFC3), thus making possible comparisons over large time baselines. While the field of view of HST is small compared to what is available from the ground, HST's ability to observe very faint sources yields more PM measurements per fixed area. The number of measurable stars in an HST field can range from $N \approx 10^{1-2}$ for a sparse halo pointing (Deason *et al.* 2012; Sohn *et al.* 2014) to $N \approx 10^{5-6}$ for the center of a globular cluster (e.g., Anderson & van der Marel 2010).

Due to HST's unparalleled astrometric capabilities, it can achieve the PM accuracy needed for LG science *for individual stars* (50 μas/yr corresponds to a motion of ~ 0.01 HST CCD pixels over 10 years). So it is possible not only to measure very accurate

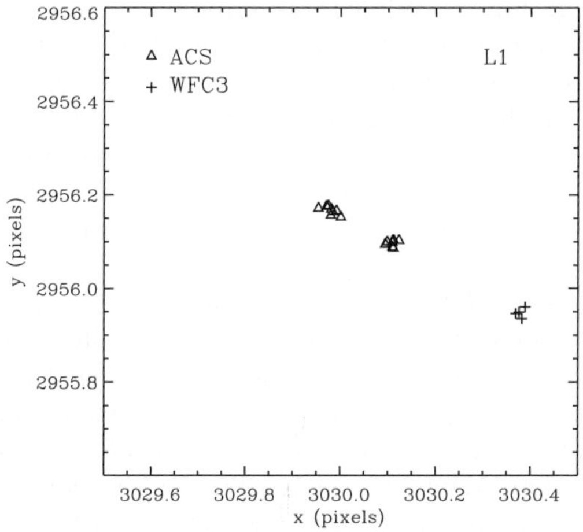

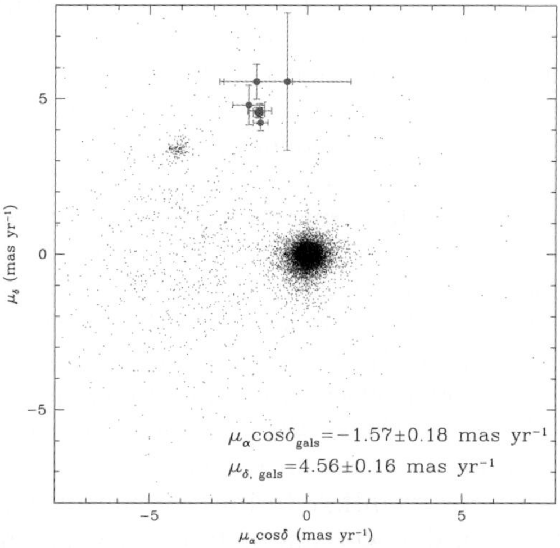

Figure 1. (left) HST measurements of the position of a background quasar relative to the stars in an LMC field (from Kallivayalil *et al.* 2013). The box is the size of a single CCD pixel. Measurements are shown with different instruments, at three epochs spread over a 7 year period. The average absolute PM of the LMC stars can be accurately measured from such data.

Figure 2 (right) PMs of all compact sources in an HST pointing centered on the bulge globular cluster M70 (from Massari *et al.* 2013), measured relative to the cluster PM. Point sources (black) are stars in either M70 (middle clump), the Sagittarius dSph galaxy in the background (top left clump), or the MW disk and bulge (extended distribution left of center). Red points with errorbars, and their weighted average in blue, are distant background galaxies which define the absolute PM zeropoint. Such data allow the measurement of both absolute bulk and relative internal motions of globular clusters and dwarf galaxies.

average PMs, but also to study internal PM kinematics of stellar systems. This includes studies of the velocity dispersion of globular clusters (not discussed in this review, but see, e.g.: McNamara *et al.* 2003, 2012; McLaughlin *et al.* 2006; van der Marel & Anderson 2010; Bellini *et al.* 2013), the rotation of galaxies (van der Marel & Kallivayalil 2014), or different kinematical populations in tidal streams (Sohn *et al.* 2014). These possibilities have led several groups over the past decade to focus on the use of HST for LG PM studies. Specifically, many of the results discussed in the following sections were obtained in the context of our HSTPROMO (Hubble Space Telescope Proper Motion) collaboration, a set of several dozen HST investigations with both observational and theoretical components (see the review in van der Marel 2014, or the HSTPROMO web page†).

Observationally determined PMs can be of two kinds: relative PMs or absolute PMs. Relative PMs are easier to obtain, and suffice e.g. if the goal is to measure the *internal* kinematics of a stellar system. But for LG science it is generally necessary to obtain absolute PMs, by measuring the motion of foreground stars with respect to distant (stationary) background objects. Most early studies (e.g., Piatek *et al.* 2002; Kallivayalil *et al.* 2006a) focused on fields centered on a single background quasar (see, e.g., Fig. 1). A quasar has the advantage of being a point source, so that its position can be measured with the same techniques as the foreground stars (Anderson & King 2000). However, this limits the applicability to the small fraction of fields that contain a known quasar. It also limits the accuracy per field to the accuracy with which a single source can be centroided.

† http://www.stsci.edu/∼marel/hstpromo.html

It is possible to achieve higher absolute PM accuracies by using compact background galaxies as stationary reference sources (see, e.g., Fig. 2). While their individual positional accuracy is somewhat poorer that for a point source, their much larger number provides an overwhelming $\sqrt{N}$ advantage in averaging (a typical 1 hour HST pointing will contain of order 10^2 background galaxies sufficiently bright and compact for astrometry). Also, background galaxies sample the whole detector, which reduces systematic errors upon averaging. Measuring positions of background galaxies is more complicated than for point sources, but sophisticated techniques for this now do exist (e.g., Mahmud & Anderson 2008; Sohn *et al.* 2012,2013,2014).

Systematic errors are often the limiting factor in PM studies. Thus it is necessary to perform careful calibrations of PSF shapes, geometric distortions, charge-transfer efficiency, color effects, time variability in calibrations, etc. (e.g., Anderson & King 2000; Sohn *et al.* 2012). When this is done, the systematics can be controlled to levels below the typical random errors. This has been confirmed for several studies by the good agreements between HST PM measurements and independent estimates from combinations of theoretical models and LOS velocity measurements (e.g., van der Marel *et al.* 2012a; van der Marel & Kallivayalil 2014; Sohn *et al.* 2014).

The results from GAIA, a full-sky astrometric satellite launched in 2013, will further revolutionize our understanding of PM dynamics. However, the best results will likely be limited to the MW and its satellites. This is because GAIA is not optimized for faint targets or crowded fields. For example, the predicted end-of-mission accuracy at magnitude $V \sim 20$ is Δ PM $\sim 430\,\mu$as/yr (this includes the contributions from scattered light discovered post-launch). By comparison, HST already reaches Δ PM $\sim 100\,\mu$as/yr at $V \sim 25$ (e.g., Sohn *et al.* 2014). Therefore, HST will continue to be a unique and complementary resource in the GAIA era. Looking further into the 2020s, the EUCLID and WFIRST-AFTA space missions will enable HST-quality PM measurements over much wider fields of view, thus opening the door for even more ambitious scientific investigations of the LG.

3. Milky Way Populations

Studies of the dynamics of the MW disk through PM measurements of stars in the solar neighborhood date back more than a century. The Hipparcos mission has produced the most important catalog in this field (e.g., Dehnen & Binney 1998), and the GAIA mission will further improve our knowledge. I will not review the local MW disk dynamics here, but will focus instead on more distant MW components and tracers.

Globular Clusters The absolute PMs of many globular clusters have been measured from ground-based observations in a series of papers by Casetti-Dinescu *et al.* (e.g., 2007, 2010, 2013). A catalog with the PMs of ~ 60 clusters is available online†. The PMs have been used, e.g., to assess the three-dimensional orbits of clusters, the equilibrium dynamics of the cluster population as a whole, the dynamical differences between disk and halo clusters, the possible external origin of some clusters, or potential associations of specific clusters with tidal streams. However, the available ground-based PM uncertainties are such that for $D \gtrsim 10\,$kpc, the corresponding velocity uncertainties are generally too large to strongly constrain these questions. More accurate HST PM measurements of globular clusters are now starting to become available as well (e.g., Massari *et al.* 2013; Sohn *et al.* 2014; see Fig. 2). These have the potential to probe the cluster dynamics to much larger distances.

† http://www.astro.yale.edu/dana/gc.html

Galactic Bulge Moving on to individual stars outside the solar neighborhood, several groups have studied the dynamics of the MW Bulge. Early ground-based PM measurements (Spaenhauer *et al.* 1992) found the velocity dispersions to be near-isotropic. Subsequent HST studies have revealed more complex kinematical signatures, such as rotation, and radial (depth) variations within the bulge (Kuijken & Rich 2002; Clarkson *et al.* 2008; Soto *et al.* 2014). These have been used to improve our understanding of the bulge's structure and its connection to the central bar (Rattenbury *et al.* 2007; Poleski *et al.* 2013).

Metal-Poor Halo The PMs of distant stars in the metal-poor halo can also be measured with HST. Deason *et al.* (2013) showed that it is possible to uniquely identify main-sequence halo stars in random, multiply-imaged HST fields, by combining PM and color-magnitude diagram (CMD) information. They measured the PMs of 13 stars at 24 ± 6 kpc toward the Andromeda-Triangulum halo overdensity, which imply a halo velocity ellipsoid that is roughly isotropic. This is more tangential than the velocity ellipsoid of halo stars near the Sun, which is measured from ground-based PMs to be highly radially anisotropic (e.g., Bond *et al.* 2010). These results are consistent with the hypothesis that the Andromeda-Triangulum overdensity may be a shell in the halo, possibly resulting from an ancient accretion event. Work is ongoing to use data for many more fields in the HST Archive to measure PMs for ~ 1000 distant stars throughout the halo (Sohn *et al.*, in prep.). This will help constrain halo formation mechanisms (through the implied velocity dispersion anisotropy of halo stars, and the amount of substructure in velocity space), and it will improve our knowledge of the radial MW mass profile by resolving the mass vs. velocity dispersion anisotropy degeneracy.

Hypervelocity Stars Dynamical interactions near the Milky Way's central supermassive black hole can eject stars at high velocities. Such stars are observed as hypervelocity stars in the MW halo (Brown *et al.* 2005). Given their distances, only HST has so far been able to measure their PMs. An accurate PM can confirm the origin near the Galactic Center (Brown *et al.* 2010), while the implied deviation from purely radial motion can in principle constrain the shape of the MW's dark halo (Gnedin *et al.* 2005).

Tidal Streams The MW halo contains many stellar streams, composed of the tidally stripped material from globular clusters or dwarf satellites. These streams hold great promise to constrain the shape and density profile of the MW's dark matter halo, especially when PMs are available in addition to sky positions, distances, and LOS velocities. Koposov *et al.* (2010) used ground-based USNO-B astrometry to study the PMs of the GD1 Stream. However, since this stream is at $D \approx 10$ kpc, it provides only limited information on the MW's dark matter halo. The Sagittarius Stream, composed of material stripped from the Sgr dSph, is a more useful tracer in this regard, since it reaches distances up to $\sim$ 100kpc (Belokurov *et al.* 2014). For the Sgr dSph itself, both ground-based (Dinescu *et al.* 2005) and HST-based PMs (Pryor *et al.* 2010; Massari *et al.* 2013; Sohn *et al.* 2014; see Fig. 2) are available. For the Sgr Stream, Carlin *et al.* (2012) and Koposov *et al.* (2013) obtained ground-based PMs along the trailing arm, while Sohn *et al.* (2014) obtained higher accuracy HST PMs along both arms. The latter study was even able to kinematically separate trailing-arm from leading-arm stars *within the same field*, based on their PMs. The average stream PMs are close to the predictions of the best-fitting Stream model previously constructed by Law & Majewski (2010). In this model, the MW dark halo is near-oblate and oriented orthogonal to the MW disk. Further measurements and models will likely be required to either validate or refute this model (see also, Deg & Widrow 2013; Ibata *et al.* 2013; Vera-Ciro & Helmi 2013). The PMs of other streams, including the Orphan stream, are also being studied with HST (van der Marel *et al.*, in prep.).

4. Dwarf Satellite Galaxies

The PMs of almost all of the MWs classical dSph satellites have been measured using HST (Piatek *et al.* 2003, 2005, 2006, 2007; Lepine *et al.* 2011; Pryor *et al.* 2014), and these are generally the most accurate measurements available. Knowledge of the PMs has been used to determine the dSph orbits. These can be used to investigate the origin of these galaxies, the cause of structural of star formation history features, and/or their connection with other satellites. For example, there has long been evidence that some MW satellites lie in a plane (e.g., Lynden-Bell & Lynden-Bell 1995). By measuring PMs it is possible to verify that the orbital angular momentum vectors are roughly perpendicular to this plane (e.g., Pawlowski & Kroupa 2013), as expected for a long-lived configuration. Also, the satellite orbits can be modeled as an ensemble under the assumption of hydrostatic equilibrium, to constrain the MW mass profile (e.g., Watkins *et al.* 2010).

Sohn *et al.* (2013) measured the absolute PM of Leo I. This is one of the most unusual MW satellites, because of its large Galactocentric distance ($r = 261 \pm 13$ kpc) and the fact that it is rapidly receding away from the MW ($v_{\rm rad} = 168 \pm 3\,{\rm km\,s^{-1}}$). The measured PM implies a significant tangential velocity $v_{\rm tan} = 101 \pm 34\ {\rm km\,s^{-1}}$, indicating that it is not on a radial orbit due to a slingshot interaction with another satellite. Orbit integrations indicate instead that it likely entered the MW virial radius for the first time only 2.3 ± 0.2 Gyr ago, and had its first pericenter passage 1.0 ± 0.1 Gyr ago at 91 ± 36 kpc. Comparison of the properties of Leo I to the properties of subhalos in cosmological simulations yields important new constraints on the mass of the MW (Boylan-Kolchin *et al.* 2013). Assuming that Leo I is the least-bound classical MW satellite, the implied MW virial mass is $M_{\rm vir} \approx (1.6 \pm 0.3) \times 10^{12}\ M_{\odot}$; for lower MW masses it would be vanishingly unlikely to find a subhalo as distant and rapidly moving as Leo I. HST studies of the PMs of other distant dwarf galaxies are currently ongoing, including both distant MW satellites such as Leo T (Do *et al.*, in prep.), as well as distant LG satellites such Cetus, Tucana, Leo A, and the Sgr dwarf Irregular galaxy (Sohn *et al.*, in prep.).

Future studies, either with HST or future facilities, hold the promise to determine also the *internal* kinematics of dwarf galaxies. Such measurements have the potential to break the mass vs. velocity dispersion anisotropy degeneracy. This can help constrain the nature of dark matter, by determining whether dark halo mass profiles have central cores or cusps (Strigari *et al.* 2007).

5. The Magellanic Clouds

One of the first crude measurements of the PMs of the Magellanic Clouds was obtained by the Hipparcos satellite (Kroupa & Bastian 1997). With HST it has been possible to significantly refine the accuracy of these measurements. Kallivayalil *et al.* (2006a,b) presented PMs based on two epochs of HST data with a 2-year time baseline. These measurements were recently refined by Kallivayalil *et al.* (2013) using a third-epoch of HST data that extends the time baseline to 7 years (see Fig. 1). The key finding from this work has been that the Magellanic Clouds move faster about the Milky Way than was previously believed based on models of the Magellanic Stream. In these models, the Clouds were assumed to have been long-term MW satellites moving on near-circular orbits in a logarithmic potential (e.g., Gardiner & Noguchi 1996). By contrast, orbit calculations based on the HST PMs in a cosmologically motivated MW halo potential

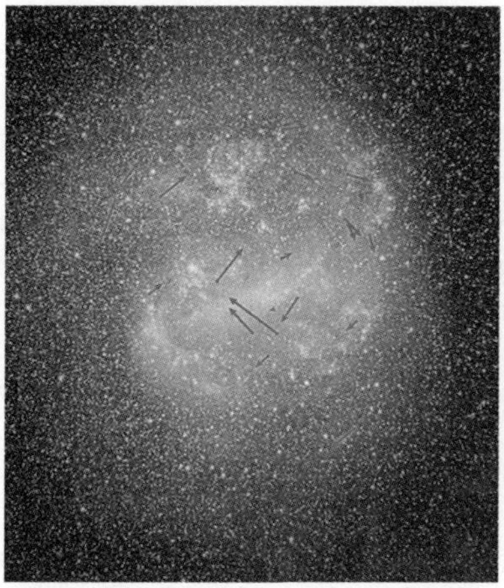

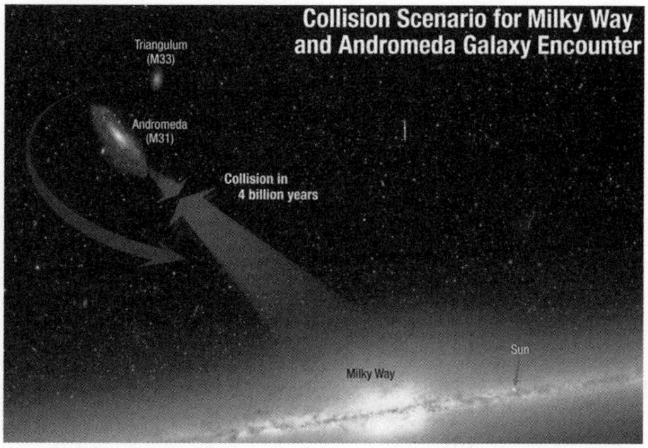

Figure 3. (left) Deep ground-based LMC image (courtesy Y. Beletsky, vertical scale $\sim 14°$), with HST measurements of the LMC PM rotation overlaid (red arrows; from van der Marel & Kallivayalil 2014). The LMC makes one revolution every few-hundred million years. The PM data can measure this rotation accurately, which (when combined with LOS data) provides a fully three-dimensional probe of the internal disk kinematics.

Figure 4 (right) Illustration of the predicted future evolution of the LG, based on the observed M31 PM from HST (Sohn *et al.* 2012) and the observed M33 PM from water maser observations (Brunthaler *et al.* 2005). The MW and M31 will collide in ~ 4 Gyr, and they will fully merge in ~ 6 Gyr. M33 will swing around M31 and take part in the collision as well (from van der Marel *et al.* 2012b).

imply either a much longer-period more-elliptical orbit, or that the Clouds are on their first infall into the MW (Besla *et al.* 2007). Given the individual LMC and SMC PM measurements, the first-infall models are the only ones in which the LMC and SMC form a bound pair (Kallivayalil *et al.* 2013), consistent with their structures and star formation histories. The plausibility of a first-infall scenario is further supported by both cosmological simulations (Boylan-Kolchin *et al.* 2011) and the gas-rich nature of the Clouds (van den Bergh 2006). In a first-infall scenario, the Magellanic Stream may have formed purely from interaction between the Clouds, without assistance from MW tidal or ram pressure forces (Besla *et al.* 2010; Peebles & Tully 2013). These results have not just refined our understanding of the Magellanic Clouds, but also of the formation of Magellanic Irregular galaxies in general (Besla *et al.* 2012).

Almost all of our knowledge of the rotation of galaxies is based on LOS velocity measurements. However, PM measurements can also probe this rotation. By mapping variations in the HST PM measurements across the face of the LMC (Kallivayalil *et al.* 2006a; Piatek *et al.* 2008), it has been possible to measure PM rotation for the first time for any galaxy. van der Marel & Kallivayalil (2014; see Fig. 3) measured both the PM rotation field and rotation curve of the LMC in detail, and demonstrated that the results are consistent with knowledge from LOS velocity measurements for individual stars. They used the three-dimensional dynamical information obtained from combination of the different techniques to obtain new insights into the geometry, structure, and distance of the LMC. HST studies of the internal PM dynamics of the SMC (Kallivayalil *et al.*, in prep.) and the Magellanic Bridge (van der Marel *et al.*, in prep.) are currently ongoing.

6. The Andromeda System

M31 was the first galaxy for which a LOS velocity was measured, and it has been known for about a century that it is moving towards the MW (Slipher 1913). However, to understand the future orbit and evolution of the M31-MW system, and hence the LG, it is necessary to also know the PM of M31. Sohn *et al.* (2012) used HST to obtain the first-ever PM measurement for M31, using data with 5–7 year time baselines for three different fields studied with multiple HST cameras. The result agrees with indirect estimates based on the LOS kinematics of M31 and LG satellites (van der Marel & Guhathakurta 2008). The combined results imply that the M31 velocity vector is statistically consistent with a radial (head-on collision) orbit towards the Milky Way ($V_{\rm tan,M31} \leqslant 34.3\,{\rm km\,s^{-1}}$ at 1σ confidence; van der Marel *et al.* 2012a).

The M31 and MW protogalaxies started moving away from each other after the Big-Bang, but their mutual gravity caused the turn-around that led to their current approach. Their combined mass implied by this "LG timing argument", including corrections for cosmic bias and scatter, is $M_{\rm LG} = M_{\rm MW,vir} + M_{\rm M31,vir} = (4.93 \pm 1.63) \times 10^{12}\,M_\odot$. Bayesian combination with other mass estimates for the individual MW and M31 galaxies yields the more accurate estimate $M_{\rm LG} = (3.17 \pm 0.57) \times 10^{12}\,M_\odot$ (van der Marel *et al.* 2012a).

N-body and Monte-Carlo simulations can be used to predict the future evolution of the MW-M31 system (van der Marel *et al.* 2012b; see Fig. 4). The galaxies will merge $t = 5.86^{+1.61}_{-0.72}$ Gyr from now. The first pericenter occurs at $t = 3.87^{+0.42}_{-0.32}$ Gyr, at a pericenter distance $r = 31.0^{+38.0}_{-19.8}$ kpc. In 41% of Monte-Carlo orbits M31 makes a direct hit with the MW, defined here as a first-pericenter distance less than 25 kpc. The PM of the Triangulum galaxy, M33, is known from water maser observations (Brunthaler *et al.* 2005), so its orbit can be calculated as well. The most likely outcome is for the MW and M31 to merge first, with M33 settling onto an orbit around them that may decay towards a merger later. However, there is a 9% probability that M33 makes a direct hit with the MW at its first pericenter, *before* M31 gets to or collides with the MW. Also, there is a 7% probability that M33 gets ejected from the Local Group, temporarily or permanently. The radial mass profile of the MW-M31 merger remnant is significantly more extended than the original profiles of either the MW or M31, and suggests that the merger remnant will resemble an elliptical galaxy. The Sun will most likely ($\sim 85\%$ probability) end up at larger radius from the center of the MW-M31 merger remnant than its current distance from the MW center, possibly further than 50 kpc ($\sim 10\%$ probability). There is a $\sim 20\%$ probability that the Sun will at some time in the next 10 Gyr find itself moving through M33 (within 10 kpc), but while dynamically still bound to the MW-M31 merger remnant.

7. Concluding Remarks

The ability to measure the PMs of stars and galaxies in the nearby Universe with HST and other facilities has led to a revolution in our understanding of the dynamics of the LG and its galaxies. The LG continues to evolve, with Leo I, and probably also the Magellanic Clouds, just now falling into the MW for the first time. In a few billion years, M31 will arrive as well, which will cause the Local Group to be drastically reshaped. With the continuing improvements in observational facilities, and the prospect of GAIA and other future space telescopes, it is likely that PM measurements will continue to improve our understanding of our dynamic LG environment. In turn, this will be a key ingredient for understanding the formation and evolution of galaxies and galaxy groups in general.

Acknowledgements

Support for this work and for HSTPROMO projects was provided by NASA through grants from STScI, which is operated by AURA, Inc., under NASA contract NAS 5-26555.

References

Anderson, J. & King, I. R. 2000, *PASP*, 112, 1360
Anderson, J. & van der Marel, R. P. 2010, *ApJ*, 710, 1032
Bellini, A., van der Marel, R. P., & Anderson, J. 2013, *Memorie della Societa Astronomica Italiana*, 84, 140
Belokurov, V., *et al.* 2014, *MNRAS*, 437, 116
Besla, G., Kallivayalil, N., Hernquist, L., Robertson, B., Cox, T. J., van der Marel, R. P., & Alcock, C. 2007, *ApJ*, 668, 949
Besla, G., Kallivayalil, N., Hernquist, L., van der Marel, R. P., Cox, T. J., & Kereš, D. 2010, *ApJL*, 721, L97
Besla, G., Kallivayalil, N., Hernquist, L., van der Marel, R. P., Cox, T. J., & Kereš, D. 2012, *MNRAS*, 421, 2109
Binney, J. & Mamon, G. A. 1982, *MNRAS*, 200, 361
Bond, N. A., *et al.* 2010, *ApJ*, 716, 1
Boylan-Kolchin, M., Besla, G., & Hernquist, L. 2011, *MNRAS*, 414, 1560
Boylan-Kolchin, M., Bullock, J. S., Sohn, S. T., Besla, G., & van der Marel, R. P. 2013, *ApJ*, 768, 140
Brown, W. R., Geller, M. J., Kenyon, S. J., & Kurtz, M. J. 2005, *ApJ*, 622, L33
Brown, W. R., Anderson, J., Gnedin, O. Y., Bond, H. E., Geller, M. J., Kenyon, S. J., & Livio, M. 2010, *ApJ*, 719, L23
Brunthaler, A., Reid, M. J., Falcke, H., Greenhill, L. J., & Henkel, C. 2005, *Science*, 307, 1440
Brunthaler, A., Reid, M. J., Falcke, H., Henkel, C., & Menton, K. M. 2007, *A&A*, 462, 101
Carlin, J. L., Majewski, S. R., Casetti-Dinescu, D. I., Law, D. R., Girard, T. M., & Patterson, R. J. 2012, *ApJ*, 744, 25
Casetti-Dinescu, D. I., Girard, T. M., Herrera, D., van Altena, W. F., Lopez, C. E., & Castillo, D. J. 2007, *AJ*, 134, 195
Casetti-Dinescu, D. I., Girard, T. M., Korchagin, V. I., van Altena, W. F., & Lopez, C. E. 2010, *AJ*, 140, 1282
Casetti-Dinescu, D. I., Girard, T. M., Jilkova, L., van Altena, W. F., Podesta, F., & Lopez, C. E. 2013, *AJ*, 146, 33
Clarkson, W., *et al.* 2008, *ApJ*, 684, 1110
Darling, J. 2011, *ApJ*, 732, L2
Deason, A. J., Van der Marel, R. P., Guhathakurta, P., Sohn, S. T., & Brown, T. M. 2013, *ApJ*, 766, 24
Deg, N. & Widrow, L. 2013, *MNRAS*, 428, 912
Dehnen, W. & Binney, J. J. 1998, *MNRAS*, 298, 387
Dinescu, D. I., Girard, T. M., van Altena, W. F., & Lopez, C. E. 2005, *ApJ*, 618, L25
Gardiner, L. T. & Noguchi, M. 1996, *MNRAS*, 278, 191
Ghez, A. M., *et al.* 2008, *ApJ*, 689, 1044
Gnedin, O. Y., Gould, A., Miralda-Escude, J., & Zentner, A. R. 2005, *ApJ*, 622, L33
Ibata, R., Lewis, G. F., Martin, N. F., Bellazzini, M., & Correnti, M. 2013, *ApJ*, 765, L15
Kallivayalil, N., van der Marel, R. P., & Alcock, C. 2006a, *ApJ*, 652, 1213
Kallivayalil, N., van der Marel, R. P., Alcock, C., Axelrod, T., Cook, K. H., Drake, A. J., & Geha, M. 2006b, *ApJ*, 638, 772
Kallivayalil, N., van der Marel, R., Besla, G., Anderson, J., & Alcock, C. 2013, *ApJ*, 764, 161
Koposov, S. E., Rix, H.-W., & Hogg, D. W. 2010, *ApJ*, 712, 260
Koposov, S. E., Belokurov, V., & Evans, W. N. 2013, *ApJ*, 766, 79
Kroupa, P. & Bastian, U. 1997, *New Astronomy*, 2, 77

Kuijken, K. & Rich, R. M. 2002, *AJ*, 124, 2054
Law, D. R. & Majewski, S. R. 2010, *ApJ*, 714, 229
Lepine, S., Koch, A., Rich, R. M., & Kuijken, K. 2011, *ApJ*, 741, 100
Lynden-Bell, D. & Lynden-Bell, R. M. 1995, *MNRAS*, 275, 429
Mahmud, N. & Anderson, J. 2008, *PASP*, 120, 907
Massari, D., Bellini, A., Ferraro, F. R., van der Marel, R. P., Anderson, J., Dalessandro, E., & Lanzoni, B. 2013, *ApJ*, in press
McLaughlin, D. E., Anderson, J., Meylan, G., Gebhardt, K., Pryor, C., Minniti, D., & Phinney, S. 2006, *ApJS*, 166, 249
McNamara, B. J., Harrison, T. E., & Anderson, J. 2003, *ApJ*, 595, 187
McNamara, B. J., Harrison, T. E., Baumgardt, H., & Khalaj, P. 2012, *ApJ*, 745, 175
Pawlowski, M. S. & Kroupa, P. 2013, *MNRAS*, 435, 2116
Peebles, P. J. E. & Tully, R. B. 2013, *ApJ*, 778, 137
Piatek, S., *et al.* 2002, *AJ*, 124, 3198
Piatek, S., Pryor, C., Olszewski, E. W., Harris, H. C., Mateo, M., Minniti, D., & Tinney, C. G. 2003, *AJ*, 126, 2346
Piatek, S., Pryor, C., Bristow, P., Olszewski, E. W., Harris, H. C., Mateo, M., Minniti, D., & Tinney, C. G. 2005, *AJ*, 130, 95
Piatek, S., Pryor, C., Bristow, P., Olszewski, E. W., Harris, H. C., Mateo, M., Minniti, D., & Tinney, C. G. 2006, *AJ*, 131, 1445
Piatek, S., Pryor, C., Bristow, P., Olszewski, E. W., Harris, H. C., Mateo, M., Minniti, D., & Tinney, C. G. 2007, *AJ*, 133, 818
Piatek, S., Pryor, C., & Olszewski, E. W. 2008, *AJ*, 135, 1024
Poleski, R., *et al.* 2013, *ApJ*, 776, 76
Pryor, C., Piatek, S., & Olszewski, E. W. 2014, *AJ*, submitted [arXiv1407.3509]
Rattenbury, N. J., Mao, S., Debattista, V. P., Sumi, T., Gerhard, O., & de Lorenzi, F. 2007, *MNRAS*, 378, 1165
Slipher, V. M. 1913, *Lowell Observatory Bulletin*, vol. 1, 56
Sohn, S. T., Anderson, J., & van der Marel, R. P. 2012, *ApJ*, 753, 7
Sohn, S. T., Besla, G., van der Marel, R. P., Boylan-Kolchin, M., Majewski, S. R., & Bullock, J. S. 2013, *ApJ*, 768, 139
Sohn, S. T., van der Marel, R. P., Carlin, J. L., Majewski, S. R., Kallivayalil, N., Law, D. R., Anderson, J., & Siegel, M. H. 2014, *ApJ*, submitted [arXiv:1408.3408]
Soto, M., Zeballos, H., Kuijken, K., Rich, R. M., Kunder, A., & Astraatmadja, T. 2014, *A&A*, 562, 41
Spaenhauer, A., Jones, B. F., & Whitford, A. E. 1992, *AJ*, 103, 297
Strigari, L. E., Bullock, J. S., & Kaplinghat, M. 2007, *ApJ*, 657, L1
van den Bergh, S. 2006, *AJ*, 132, 1571
van der Marel, R. P. & Guhathakurta, P. 2008, *ApJ*, 678, 187
van der Marel, R. P. & Anderson, J. 2010, *ApJ*, 710, 1063
van der Marel, R. P., Fardal, M., Besla, G., Beaton, R. L., Sohn, S. T., Anderson, J., Brown, T., & Guhathakurta, P. 2012a, *ApJ*, 753, 8
van der Marel, R. P., Besla, G., Cox, T. J., Sohn, S. T., & Anderson, J. 2012b, *ApJ*, 753, 9
van der Marel, R. P. & Kallivayalil, N. 2014, *ApJ*, 781, 121
van der Marel, R. P., *et al.* 2014, in *Proc. of Structure and Dynamics of Disk Galaxies*, M. S. Seigar, & P. Treuthardt, eds., ASP Conference Series, Vol. 480 , 2014, p.43
Vera-Ciro, C. & Helmi, A. 2013, *ApJ*, 773, L4
Watkins, L. L., Evans, N. W., & An, J. H. 2010, *MNRAS*, 406, 264

Galaxy Masses as Constraints of Formation Models
Proceedings IAU Symposium No. 311, 2014
M. Cappellari & S. Courteau, eds.

doi:10.1017/S1743921315003300

Mass modelling from stellar streams in the Milky Way

Amina Helmi[1] **and Robyn E. Sanderson**[1,2,3]

[1]Kapteyn Astronomical Institute, University of Groningen, P.O.Box 800, 9700 AV Groningen, The Netherlands. email: ahelmi@astro.rug.nl
[2]Columbia University Department of Astronomy, Mail Code 5246, 550 W 120th St, New York, NY 10027. email: robyn@astro.columbia.edu
[3]NSF Astronomy & Astrophysics Postdoctoral Fellow

1. Introduction

Arguably two of the most important questions in Astrophysics today are: what is the Universe made of? and, how do galaxies form and evolve? Quite astonishingly we know only the properties of $< 5\%$ of the mass in the Universe (the atoms we are made of), while the nature of the dominant mass component (the dark matter) remains elusive. However, because of its abundance, dark matter is a significant driver of both the dynamics of stars and galaxies and of their formation and evolution.

N-body simulations of the formation of structure in the Universe make very definite predictions for the distribution of mass, the shape, the time-evolution and the granularity of dark matter halos. These predictions depend strongly on the fundamental nature of the dark matter (that is, on the properties of the dark matter particles themselves). For example, if the dark matter is cold, the halos are triaxial in shape (Vera-Ciro *et al.* 2011), and follow specific density profiles that behave as $1/r$ near the centres and $1/r^3$ at large distances (Navarro, Frenk & White 1996). They also host myriads of small satellites or dark matter subhalos (Springel *et al.* 2008). The characteristics of the dark matter halos hosting galaxies, such as their mass and scale radius, also depend slightly on the cosmological model, as systems formed in cold dark matter are more concentrated than those in warm dark matter (Lovell *et al.* 2014). Although many of these predictions derive from simulations that do not include the effect of baryons, these are largely only dominant in the central regions of the host halos, and therefore, such predictions should be preferably tested far out, in the outer regions of galaxies.

Tidal streams are ideally suited to test these cosmological predictions on the distribution of dark matter as well as alternative theories of gravity (Famaey & Binney 2005). This is because stream stars follow nearly parallel orbits thereby mapping the gravitational field in which they move.

For the Milky Way, tests of the cosmological predictions are not only feasible now but can also be much more constraining because the number and variety of observations that need to be fitted simultaneously. And while much has been done, the results to-date are still inconclusive. For example the Milky Way's halo mass is not known to better than within a factor of 2 (100% uncertainty). There are only weak constraints on its density profile. Its shape has received a lot of attention, yet in the past ten years its has been claimed to be round, oblate, prolate, extremely triaxial and highly unusual (Law & Majewski 2010), or consistent with cosmological expectations if the effect of the Large Magellanic Cloud is considered (Vera-Ciro & Helmi 2013). The reason that progress has been limited in this particular area is that very few streams (just two!) are currently known with full and reasonably accurate phase-space information (Koposov *et al.* 2010), and this information is imperative for resolving degeneracies in the modelling and for obtaining the desired tight constraints (Sanders & Binney 2013).

However, with the launch of the Gaia mission, the situation will change dramatically in the coming years. The expectation is that Gaia will lead to the discovery of hundreds of streams (Helmi *et al.* 2011) that will have accurate phase-space coordinates. This of course calls for the use of sophisticated tools for the dynamical modelling.

2. A new method to derive the mass of the Milky Way from stellar streams

Although streams may stretch hundreds of degrees on the sky, they constitute very tight clumps in phase-space as shown in Figure 1. This is because they originate in objects that are initially compact, and this compactness is preserved in time (Liouville's theorem). The most suitable set of coordinates to describe these systems in phase-space are action-angle coordinates. These quantities are dependent on the gravitational potential, and it can be shown that the degree of clustering of streams in the space of actions is largest when these are computed for the correct gravitational potential.

Recently, Sanderson *et al.* (2014) have developed a statistical method based on this information content to show that the mass and scale radius of the Galactic dark halo could be recovered accurately with Gaia if assumed to be spherically symmetric. As we describe below, this is based on the application of the Kullback-Leibler (Kullback & Leibler 1951) divergence (KLD), which can be used for clustering estimation (i.e. to determine for which values of some characteristic parameters of the potential the clustering is maximal) and also for assigning significance levels/uncertainties to the recovered parameters.

The KLD measures the relative information content of two distributions, so we use it to measure clustering in two steps. First we choose a form for the potential and trial values for its parameters $\mathbf{a}$. We compute the actions $\mathbf{J}$ for a Mock sample of halo streams' stars (see below) and compare their action distribution, $f_{\mathbf{a}}(\mathbf{J})$, to the product of its marginal distributions $f_{\mathbf{a}}^{\mathrm{shuf}}(\mathbf{J})$, constructed by shuffling each star's three actions relative to one another. Shuffling eliminates correlations between different actions, so $f^{\mathrm{shuf}}(\mathbf{J})$ will always contain less information than $f(\mathbf{J})$. To find the best-fit potential we change the parameters of the model to maximize the KLD†

$$D_{\mathrm{KL}}^{\mathrm{I}} = \int f_{\mathbf{a}}(\mathbf{J}) \log \frac{f_{\mathbf{a}}(\mathbf{J})}{f_{\mathbf{a}}^{\mathrm{shuf}}(\mathbf{J})} \, d^3\mathbf{J}. \tag{2.1}$$

Once we identify the best fit, denoted $\mathbf{a}_0$, we use the KLD again to compare the distribution of stellar actions computed using the best fit with the distribution for other choices of parameters $\mathbf{a}_{\mathrm{trial}}$ (which should always be less clustered, i.e. less informative, than the distribution using $\mathbf{a}_0$):

$$D_{\mathrm{KL}}^{\mathrm{II}} = \int f_{\mathbf{a}_0}(\mathbf{J}) \log \frac{f_{\mathbf{a}_0}(\mathbf{J})}{f_{\mathbf{a}_{\mathrm{trial}}}(\mathbf{J})} \, d^3\mathbf{J}. \tag{2.2}$$

In Sanderson *et al.* (2014) we show that this KLD value can be interpreted as the difference in log probability of the parameters $\mathbf{a}_0$ and $\mathbf{a}_{\mathrm{trial}}$, averaged over the data sample:

$$D_{\mathrm{KL}}^{\mathrm{II}} = \left\langle \log \frac{\mathcal{P}(\mathcal{H}_{\mathbf{a}_0}|\mathbf{J})}{\mathcal{P}(\mathcal{H}_{\mathbf{a}_{\mathrm{trial}}}|\mathbf{J})} \right\rangle_{\mathbf{J}}, \tag{2.3}$$

where $\mathcal{H}_{\mathbf{a}_0}$ and $\mathcal{H}_{\mathbf{a}_{\mathrm{trial}}}$ represent the hypotheses that the distribution of $\mathbf{J}$ at the best

† In statistics literature this version of the KLD is known as the mutual information.

fit came from a potential with parameters $\mathbf{a}_0$ or $\mathbf{a}_{\rm trial}$, respectively. This allows us to set probability contours around the best-fit result. For instance, parameters for which $D_{\rm KL}^{\rm II} = 1/2$ are about 24 percent as likely as the best fit (using a base-e logarithm). In the case of a Gaussian distribution, this level of relative probability is attained at 1σ on either side of the mean value, so we set this level as an analogue to a traditional 1σ contour. Likewise $D_{\rm KL}^{\rm II} = 2$ and $D_{\rm KL}^{\rm II} = 9/2$ are roughly analogous to traditional 2- and 3σ contours.

To test the method we construct a toy model of the stellar halo built up by 150 satellites, initially set up as Plummer spheres with mass and luminosity distributions similar to those for present-day MW satellites. We integrate the satellite's (test) particles in an isochrone potential with known parameters: the scale radius r_s and total mass M. For simplicity we use one particle for each red giant star ($M_V = 1$) in each satellite, rather than constructing a full stellar population model. The resulting stellar streams have a uniform range of ages (integration times) so some are more phase-mixed than others. The mock halo is convolved with the Gaia pre-launch error model for end-of-mission performance and the sample for fitting is selected by (i) requiring that all six phase-space coordinates are measured by Gaia and (ii) selecting stars with estimated energy above a cutoff value to eliminate action-space crowding deep in the potential, which interferes with the fit. The fitting sample is shown in Figure 1, on the top panel in Galactic coordinates and on the bottom in a projection of action space (computed using the input potential parameters). Even when including Gaia errors the action-space distribution remains clumpy.

Using the KLD method described above, we find the best fit and produce contours in relative probability in the space of the two potential parameters, shown in the left panel of Figure 2. We pick out the contours analogous to $1(2,3)\sigma$ in green (yellow, red). Streams are most sensitive to the mass enclosed by their orbit, so we transform the M-axis to show the enclosed mass at the mean distance of all the stars in the fitting sample (6.5 kpc) instead of total mass. This parameter is recovered very accurately, but there is a bias in the determination of the scale radius r_s due to the small distance range of the stars in the fitting sample (1-25 kpc) compared to the true scale radius (8 kpc). If we ignore observational error momentarily and include stars for which Gaia will not measure radial velocities but ground-based spectroscopic surveys could, the upper limit on distance increases to 100 kpc (the mean distance increases to 8.6 kpc) and the bias in r_s is eliminated, as shown in the right panel of Figure 2.

3. Conclusions

The recently launched Gaia mission will deliver a full phase-space catalogue for more than 10^8 stars in the Galaxy. It is expected that this dataset will allow us to discover several hundred streams from disrupted satellites (Helmi *et al.* 2011) . The dynamical modeling of these streams will lead to precise estimates of the mass distribution of the dark halo of our Galaxy. Several methods have been proposed in the literature to this end, and some of the most promising include the use of clustering and correlations in action-angle space. Here we presented a new method that uses only the action information, and that can recover the mass and scale of a spherical mass distribution with ~25% precision. With phase-space information from Gaia alone, the enclosed mass at the average radius is accurate to 3% but the measurement of the scale radius is biased; including radial velocities for faint stars from ground-based follow-up allows us to determine the *total* mass and scale radius to 10% accuracy and 20% precision. One critical advantage of the method is that it does not require the identification of the individual streams, nor the

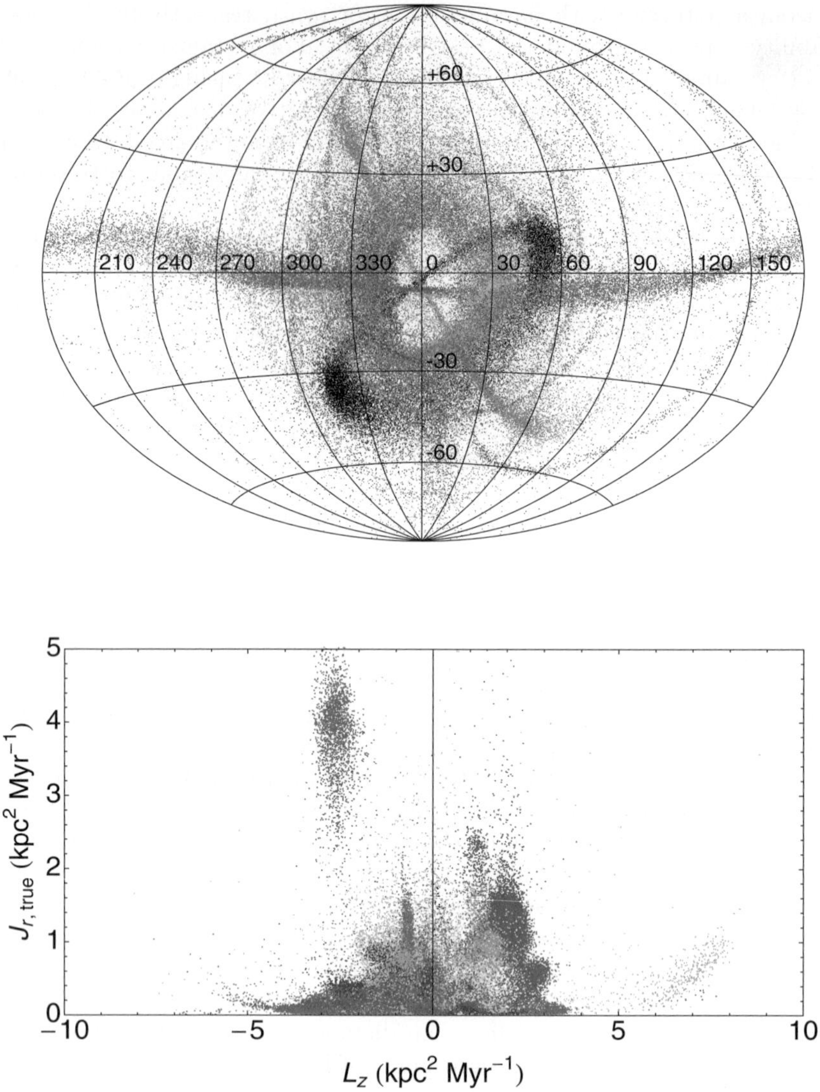

Figure 1. Tidal streams in the mock stellar halo used to test the method, in Galactic coordinates (top) and projected action space (radial action J_r versus z angular momentum L_z; bottom). Each progenitor satellite is shown in a different color.

membership of stars, and only requires that significant clustering is present in action space, as expected in the context of the concordance cosmological model (Gomez *et al.* 2013).

Acknowledgements

AH and RES acknowledge financial support from European Research Council under ERC-StG grant GALACTICA-240271. RES is supported by the National Science Foundation under Award No. AST-1400989.

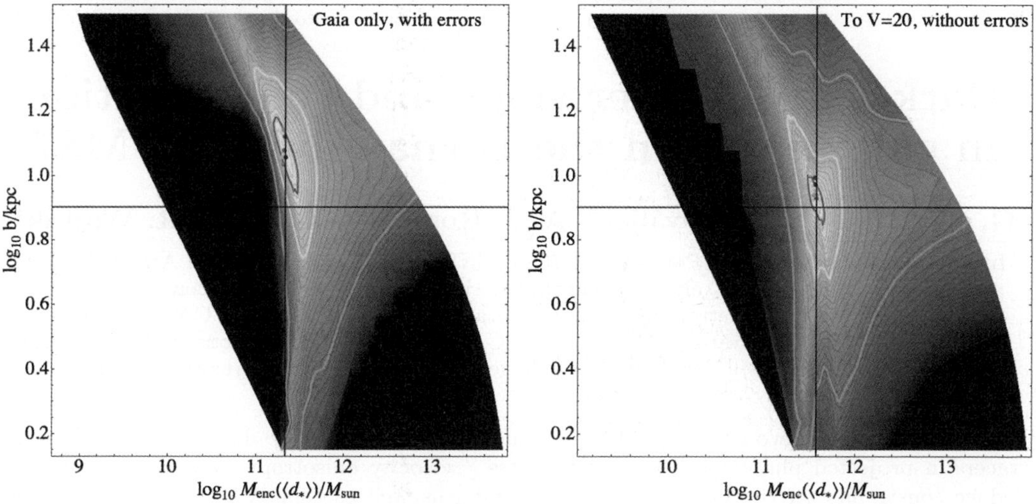

Figure 2. Contours of $D^{\rm II}_{\rm KL}$ in scale radius b and enclosed mass $M_{\rm enc}$ at the average distance of stars in the sample $\langle d_* \rangle$, for two different fitting samples. On the left are results from the error-convolved fitting sample when including only stars with Gaia radial velocities. On the right are results from a sample without errors that includes all stars for which Gaia will measure proper motions, including those whose radial velocities must be obtained by follow-up from the ground. In both panels the red (yellow, green) contours show the $D^{\rm II}_{\rm KL} = 1/2(2, 9/2)$ levels, analogous to $1(2,3)\sigma$ in the Gaussian case; the thin contours show steps of 0.2 in $D^{\rm II}_{\rm KL}$. The solid crosshairs intersect at the input values of the parameters; the colored points (red through purple) are the first through seventh highest values of $D^{\rm I}_{\rm KL}$, which is maximized to find the best-fit parameters (red).

References

Famaey, B. & Binney, J., 2005, *MNRAS*, 363, 603

Gomez, F. A., Helmi, A., *et al.* 2013, *MNRAS* 436, 3602

Helmi *et al.*, 2011, *ApJ*, 733, L7

Koposov, S. E., Rix, H.-W., & Hogg, D. W. 2010, *ApJ*, 712, 260

Kullback, S. & Leibler, R. A. 1951, *The Annals of Mathematical Statistics*, 22, 79

Law, D. R. & Majewski, S. R. 2010, *ApJ*, 714, 229

Lovell, M. R., Frenk, C. S., Eke, V. R., *et al.* 2014, *MNRAS*, 439, 300

Navarro J., Frenk C. & White S. 1996, *ApJ*, 462, 563

Sanders, J. L. & Binney, J. 2013, *MNRAS*, 433, 1826

Sanderson, R., Helmi, A., & Hogg, D. W. 2014, *Action-space clustering of tidal streams to infer the Galactic potential*, submitted to ApJ, arXiv:1404.6534

Springel V. *et al.* 2008, *MNRAS*, 391, 1685

Vera-Ciro, C. & Helmi, A. 2013, *ApJ*, 773, L4

Vera-Ciro, C. A., Sales, L. V., Helmi, A., *et al.* 2011, *MNRAS*, 416, 1377

Galaxy Masses as Constraints of Formation Models
Proceedings IAU Symposium No. 311, 2014
M. Cappellari & S. Courteau, eds.

doi:10.1017/S1743921315003312

Dark matter inner slope and concentration in galaxies: from the Fornax dwarf to M87

G. A. Mamon[1], **J. Chevalier**[1], **A. J. Romanowsky**[2,3] **and R. Wojtak**[4]

[1]Institut d'Astrophysique de Paris, UMR 7095 (CNRS & UPMC), 98 bis Bd Arago, F-75014 Paris, France, email: gam@iap.fr, jill.chevalier@gmail.com
[2]Dept. of Physics and Astronomy, San Josè State University, San Josè, CA 95192, USA
[3]Univ. of California Observatories, Santa Cruz, CA 95064, USA, email: romanow@ucolick.org
[4]Dark Cosmology Centre, DK-2100 Copenhagen, Denmark, email: wojtak@stanford.edu

Abstract. We apply two new state-of-the-art methods that model the distribution of observed tracers in projected phase space to lift the mass / velocity anisotropy (VA) degeneracy and deduce constraints on the mass profiles of galaxies, as well as their VA. We first show how a distribution function based method applied to the satellite kinematics of otherwise isolated SDSS galaxies shows convincing observational evidence of age matching: red galaxies have more concentrated dark matter (DM) halos than blue galaxies of the same stellar or halo mass. Then, applying the MAMPOSSt technique to M87 (traced by its red and blue globular clusters) we find that very cuspy DM is favored, unless we release priors on DM concentration or stellar mass (leading to unconstrained slope). For the Fornax dwarf spheroidal (traced by its metal-rich and metal-poor stars), the inner DM slope is unconstrained, with weak evidence for a core if the stellar mass is fixed. This highlights how priors are crucial for DM modeling. Finally, we find that blue GCs around M87 and metal-rich stars in Fornax have tangential outer VA.

Keywords. galaxies: elliptical and lenticular, cD - dark matter

1. Introduction

While DM cusps (with slopes steeper than $\gamma = -0.9$ for radii $r > r_{-2}/100$, i.e. $r > r_{\rm vir}/1000$) are predicted in DM-only cosmological N-body simulations (Navarro *et al.* 2004), it is not clear how the astrophysics of baryons will affect the slope γ of the inner DM density profile. Pontzen & Governato (2012) have shown that intermittent feedback can transform the DM cusps into homogeneous cores, while hydrodynamical cosmological N body simulations by Di Cintio *et al.* (2014) lead to the view that γ is a function of the stellar to halo mass ratio, where L^* ellipticals and dwarf spheroidals (dSphs) are expected to be cuspy while galaxies of intermediate mass should have DM cores.

Mass modeling of dSphs (using internal motions as the only possible method available, and now with over 1000 and 200 member velocities available for Sculptor and Fornax, respectively) leads to a variety of results. For Sculptor, Walker & Peñarrubia (2011) used the Wolf *et al.* (2010) pinch (insensitivity of the mass profile to the a priori unknown velocity anisotropy [VA] at some "pinch" radius) on metal-rich (Z-rich) vs. metal-poor (Z-poor) stars to deduce a shallow DM slope for Sculptor. However, this result has been disputed by Breddels *et al.* (2013) and Richardson & Fairbairn (2014), who both found (using orbit modeling and dispersion-kurtosis Jeans modeling, respectively) that both cores and cusps were allowed for the DM. While Jardel & Gebhardt (2012), using orbit modeling, deduced cored DM for the Fornax dSph, Amorisco *et al.* (2013) (using the Wolf pinch) found that while a DM core is the natural outcome for Fornax, (cuspy) NFW DM is possible if its concentration ($r_{\rm vir}/r_{-2}$) is low.

Similarly, two recent studies of the M87 giant elliptical (in the center of the Virgo cluster), traced by 275 to 375 globular clusters (GCs), lead to discrepant results: Murphy *et al.* (2011) (who also considered stars) deduced cored DM (with orbit modeling), while Agnello *et al.* (2014) found cuspy DM (with Jeans modeling).

2. Methods

There are two classes of methods to handle kinematic data of spherical systems (see Chapter 5 of the review of Courteau *et al.* 2014): 1) matching the line-of-sight (LOS) velocity moments predicted by the Jeans equation with those measured in bins of projected radii; 2) fitting the full distribution of observed tracers in projected phase space (PPS), using 6D distribution functions (DFs) expressed in energy and angular momentum, $f(E, J)$ (i.e. orbit modeling). Each has its weaknesses: The results with moments depend on the choice of radial bins (see Richardson & Fairbairn 2013) and do not use the full PPS information. Orbit modeling currently cannot continuously update parameters of the mass profile.

Wojtak *et al.* (2009) developed a PPS method where $f(E, J)$ is separable, taken from Wojtak *et al.* (2008), who derived the J term of the DF from first principles and checked the separable form of the DF on cluster-mass ΛCDM halos. While this method is ideal for clusters of galaxies, where dissipational effects are thought to be unimportant, it is not clear how the separability of the DF will survive the effects of dissipation expected in galaxies. Unfortunately, this method is slow, as it involves a triple integral to express the distribution of tracers in PPS in terms of the DF (Dejonghe & Merritt 1992).

Mamon, Biviano, & Boué (2013) developed a very rapid alternative, called MAMPOSSt, which is as accurate. MAMPOSSt assumes a form for the 3D velocity distribution of tracers (here, Gaussian), which allows one to predict the distribution of tracers in PPS as a single integral of the LOS velocity distribution function.

3. Age matching from galaxy mass modeling

Analyzing, with the DF method of Wojtak *et al.* (2009), the kinematics of isolated SDSS galaxies, traced by their satellites (stacked in 9 bins of host galaxy stellar mass and color), Wojtak & Mamon (2013) discovered that the DM halos of red galaxies are more concentrated ($2\,\sigma$ level) than those of blue galaxies of the same stellar or halo mass. This is a stark observational confirmation of *age matching* (Hearin & Watson 2013): halos that assemble earlier (more concentrated) lead to galaxies with older stellar populations (redder). Their trends of halo concentration vs. mass for red and blue galaxies are well matched by predictions from age matching theory (Watson, private communication).

4. M87

We trace the DM profile of M87 using GCs. We use the SLUGGS (Strader *et al.* 2011) database of GCs, removing all objects classified as UCDs, or transition objects (i.e. only keeping objects with half-mass radii lower than 5.25 pc, following Strader *et al.*), and also removing all velocities measured before 2008 (which we suspect may contain important systematic errors). This leaves us with 143 red and 322 blue GCs out to 200 kpc. Since the spectroscopic sample is incomplete, we adopt the number density profiles based on photometric samples (Table 11 of Strader *et al.*).

We ran MAMPOSSt for several models. Our standard model has a generalized NFW DM profile (see Sect. 5) with the concentration-mass ($c - M$) relation found in ΛCDM

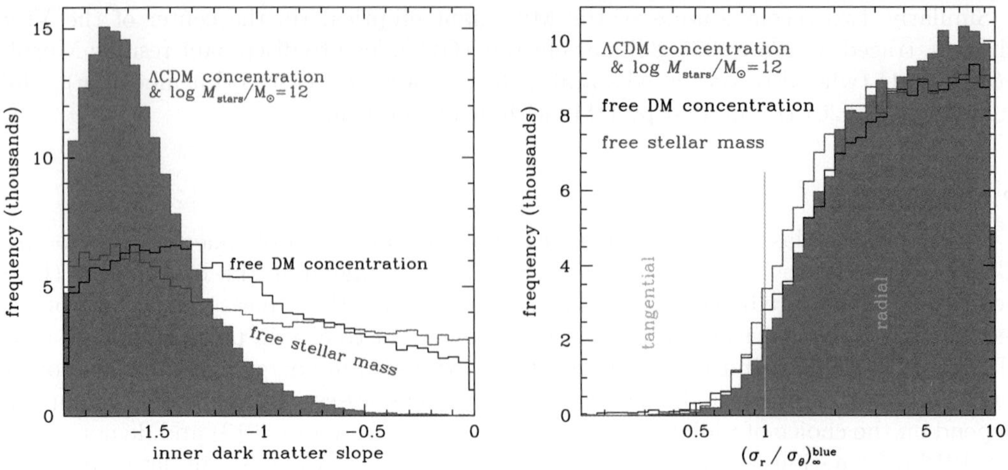

Figure 1. Effects of the choice of priors on the PDFs obtained with MAMPOSSt of the inner DM slope (*left*) and the outer VA of blue globular clusters (*right*) for M87 (traced by red and blue globular clusters). The *pink shaded regions* correspond to standard prior with $M_{\rm stars}/L_V = 8.4$ and the ΛCDM DM concentration - virial mass relation, while the other two cases are for free stellar mass (*blue*, best fit corresponding to $M_{\rm stars}/L_V = 25$, which is abnormally high) or free DM concentration (*black*, best fit at $c = 5$, a low value perhaps due to the Virgo cluster).

halos with Planck cosmology (Dutton & Macciò 2014). Its stellar mass is fixed to $10^{12} M_\odot$ (yielding $M/L_V = 8.4$). Our model also assumes Tiret *et al.* (2007) VA with isotropic inner velocities and a transition radius equal to the effective radius of the tracer.

The left panel of Figure 1 shows how γ depends crucially on the priors of the model. While in our standard model (ΛCDM concentration and fixed stellar mass, pink shaded region) $\gamma = -1.75 \pm 0.25$, significantly *cuspier* than NFW, the models with either a free concentration (black histogram) or free stellar mass (blue) allow shallower slopes, including cores, although $\gamma < -1.3$ is again preferred. Interestingly, the outer VA of the blue GCs is always very radial in all three cases (right panel of Fig. 1). This analysis can be improved by including stellar velocity dispersions and the surrounding Virgo cluster.

5. Fornax

We extracted Walker *et al.* (2009)'s spectroscopic data for the Fornax dSph, only keeping stars with over 95% confidence of not being a foreground Milky Way star (according to the database of Walker *et al.* 2009), removing velocity outliers using the cuts from Łokas (2009), splitting between Z-poor and Z-rich components using a critical value for $W' \equiv \Sigma{\rm Mg}'$ of 0.5 (i.e., Z rich corresponds to $W' > 0.5$). This yielded 722 and 1544 Z-rich and Z-poor stars, respectively. For both stellar components, we assume $n = 0.71$ Sérsic profiles with $R_e = 0.7\,{\rm kpc}$ (Battaglia *et al.* 2006). We ran MAMPOSSt on a standard model with exponentially truncated DM (as found for tidally truncated subhalos by Kazantzidis *et al.* 2004) and free inner DM slope, Tiret *et al.* (2007) VA, and the ΛCDM c–M relation, allowing the total stellar mass to be free. Figure 2-left shows that the Z-rich population displays lower velocity dispersions at small projected radii ($R < 0.3\,{\rm kpc}$).

Figure 2-right shows that the inner DM slope is not well constrained, regardless of the priors on the DM concentration and stellar mass (despite very different best fits), because if the stellar mass is free, it dominates the DM within $2\,R_e$ (with asymptotic $\log M_{\rm stars} = 8.2$). Finally, a robust finding is that the Z-rich outer VA is tangential.

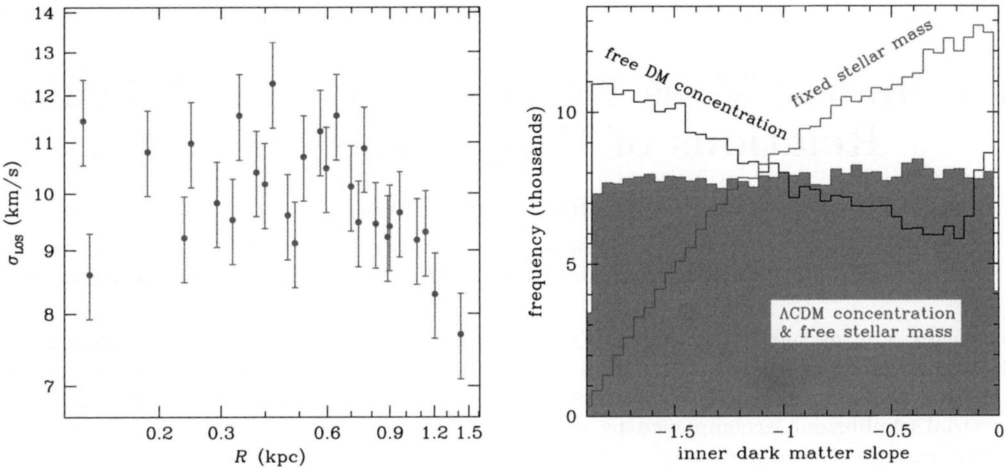

Figure 2. Analysis of the Fornax dSph. *Left*: Line-of-sight velocity dispersion profiles for the metal-rich (*red*) and metal-poor (*blue*) stars. *Right*: Effects of priors on the PDF of the inner DM slope of Fornax obtained with MAMPOSSt. The *pink shaded region* shows the standard prior with truncated outer DM, the ΛCDM DM concentration - virial mass relation, and free stellar mass, while the other 2 cases are free DM concentration (*black*, best fit at $c = 10$), or stellar mass fixed to $10^{7.95} M_\odot$ (*blue*).

References

Agnello, A., Evans, N. W., Romanowsky, A. J., & Brodie, J. P., 2014, *MNRAS*, 442, 3299
Amorisco, N. C., Agnello, A., & Evans, N. W., 2013, *MNRAS*, 429, L89
Battaglia, G. *et al.*, 2006, *A&A*, 459, 423
Breddels, M. A., *et al.*, 2013, *MNRAS*, 433, 3173
Courteau, S. *et al.*, 2014, *Reviews of Modern Physics*, 86, 47
Dejonghe, H. & Merritt, D., 1992, *ApJ*, 391, 531
Di Cintio, A., Brook, C. B., Macciò, A. V., *et al.*, 2014, *MNRAS*, 437, 415
Dutton, A. A. & Macciò, A. V., 2014, *MNRAS*, 441, 3359
Hearin, A. P. & Watson, D. F., 2013, *MNRAS*, 435, 1313
Jardel, J. R. & Gebhardt, K., 2012, *ApJ*, 746, 89
Kazantzidis, S., Mayer, L., Mastropietro, C., *et al.*, 2004, *ApJ*, 608, 663
Łokas, E. L., 2009, *MNRAS*, 394, L102
Mamon, G. A., Biviano, A., & Boué, G., 2013, *MNRAS*, 429, 3079
Murphy, J. D., Gebhardt, K., & Adams, J. J., 2011, *ApJ*, 729, 129
Navarro, J. F. *et al.*, 2004, *MNRAS*, 349, 1039
Pontzen, A. & Governato, F., 2012, *MNRAS*, 421, 3464
Richardson, T. & Fairbairn, M., 2013, *MNRAS*, 432, 3361
Richardson, T. & Fairbairn, M., 2014, *MNRAS*, 441, 1584
Strader, J. *et al.*, 2011, *ApJS*, 197, 33
Tiret, O., Combes, F., Angus, G. W., Famaey, B., & Zhao, H. S., 2007, *A&A*, 476, L1
Walker, M. G., Mateo, M., & Olszewski, E. W., 2009, *AJ*, 137, 3100
Walker, M. G. & Peñarrubia, J., 2011, *ApJ*, 742, 20
Wojtak, R., Łokas, E. L., Mamon, G. A., *et al.* 2009, *MNRAS*, 399, 812
Wojtak, R., Łokas, E. L., Mamon, G. A., *et al.*, 2008, *MNRAS*, 388, 815
Wojtak, R. & Mamon, G. A., 2013, *MNRAS*, 428, 2407
Wolf, J., Martinez, G. D., Bullock, J. S., *et al.*, 2010, *MNRAS*, 406, 1220

Galaxy Masses as Constraints of Formation Models
Proceedings IAU Symposium No. 311, 2014
M. Cappellari & S. Courteau, eds.

doi:10.1017/S1743921315003324

Dynamical Mass Determinations and Scaling Relations of Early-Type Galaxies

Michele Cappellari

Sub-department of Astrophysics, Department of Physics, University of Oxford, Denys Wilkinson Building, Keble Road, Oxford OX1 3RH. email: cappellari@astro.ox.ac.uk

Abstract. I review our understanding of classic dynamical scaling relations, relating luminosity, size and kinematics of early-type galaxies. Using unbiased determinations of galaxy mass profiles from stellar dynamical models, a simple picture has emerged in which scaling relations are driven by virial equilibrium, accompanied by a trend in the stellar mass-to-light ratio (M/L). This picture confirms the earliest insights. The trend is mainly due to the combined variation of age, metallicity and the stellar initial mass function (IMF). The systematic variations best correlate with the galaxy velocity dispersion, which traces the bulge mass fraction. This indicates a link between bulge growth and quenching of star formation. Dark matter is unimportant within the half-light radius, where the total mass profile is close to isothermal ($\rho \propto r^{-2}$).

Keywords. galaxies: elliptical and lenticular, cD - galaxies: evolution - galaxies: formation

1. Introduction

Dynamical scaling relations of early-type (elliptical E and lenticular S0) galaxies (ETGs) relate the size, luminosity (L) and stellar kinematics of galaxies. Sizes are typically described by the half-light radius ($R_{\rm e}$), while kinematics is generally quantified by the stellar velocity dispersion (σ) within a given aperture, which in this review I assume no larger than $R_{\rm e}$. Given that luminosity and size depend on distance, while kinematics do not, one of the first key applications of galaxy scaling relations was to infer galaxy distances (Dressler *et al.* 1987; Djorgovski & Davis 1987).

In more recent times, more accurate distance determination techniques have been developed, mostly based on characteristics of the galaxy stellar population (e.g. Tonry *et al.* 2001). Moreover we have a deeper understanding of the relation between redshift and distance (e.g. Planck Collaboration 2014). For these reasons distance determination is not any more the main use of dynamical scaling relations.

Instead, dynamical scaling relations are nowadays a key tool to study galaxy formation. The main reasons for this is: (i) because they provide a statistical description for easily measurable characteristics of galaxies as a function of time (redshift), which can be directly compared with numerical simulations (e.g. Robertson *et al.* 2006; Boylan Kolchin, Ma & Quataert 2006; Oser *et al.* 2012; Porter *et al.* 2014); and (ii) due to the fact that scaling parameters are actually expected to evolve very differently depending on the galaxy formation mechanism (e.g. Naab, Johansson & Ostriker 2009; Hopkins *et al.* 2010).

In this review I focus on what one can learn about dynamical scaling relations using dynamical models. Some of the results I describe are closely linked to findings made using strong lensing or stellar population synthesis approaches. These were reviewed separately at this Symposium, in particular by Tommaso Treu and Charlie Conroy, and are just briefly mentioned here. The reader is referred to Courteau *et al.* (2014) for a combined review of all these different mass determination techniques.

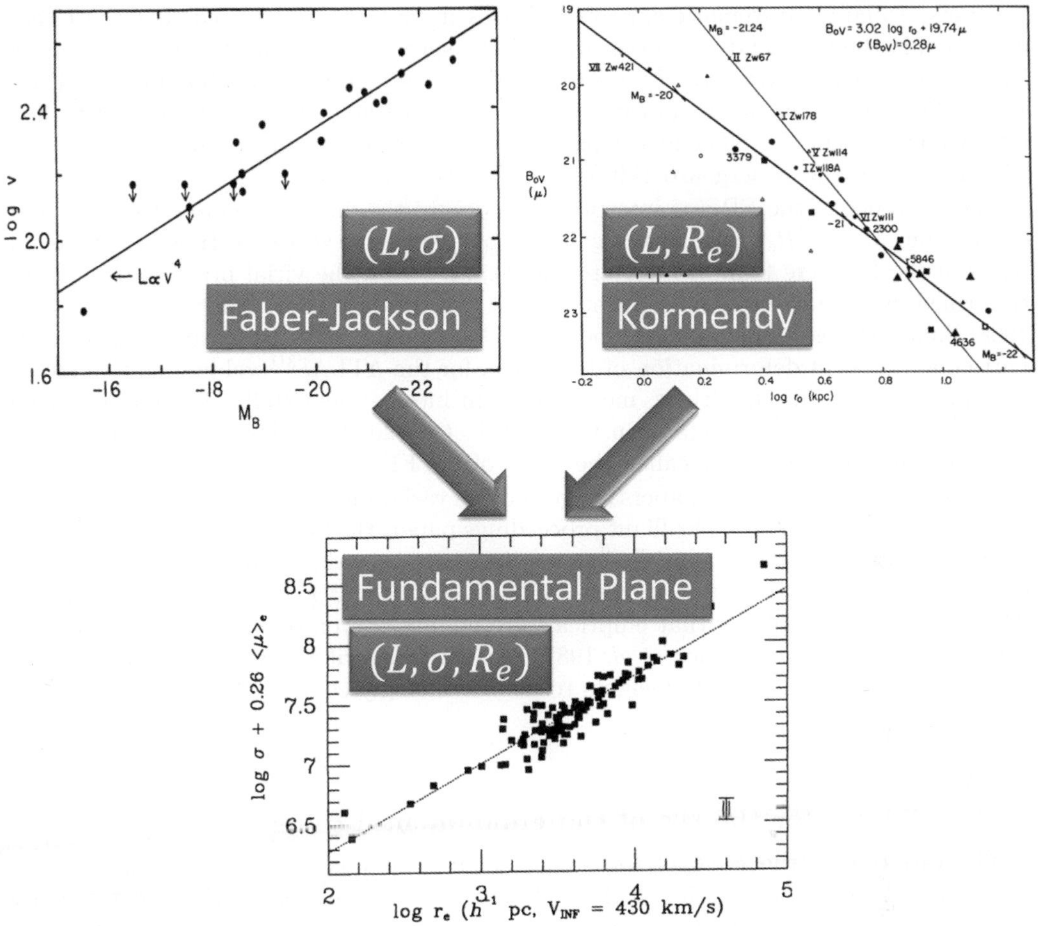

Figure 1. Classic scaling relations. The Faber-Jackson and the Kormendy relations are two special projection of a more fundamental one, aptly named the Fundamental Plane. The three figures are taken from Faber & Jackson (1976), Kormendy (1977) and Djorgovski & Davis (1987) respectively.

2. Classic scaling relations

The first dynamical scaling relation to be discovered was the one between luminosity and stellar velocity dispersion (Faber & Jackson 1976, top-left in Fig. 1). The observed relation had the form $L \propto \sigma^4$ and the authors pointed out it also suggests a trend between the mass-to-light ratio (M/L) and galaxy luminosity.

Soon thereafter, a correlation between galaxy surface brightness Σ and galaxy size was also found (Kormendy 1977, top-right in Fig. 1). When one defines the surface brightness as the mean value within $R_{\rm e}$, then $\Sigma_e = L/(2\pi R_e^2)$. This means that the Kormendy relation describes a correlation between galaxy radius and luminosity. This form has the advantage that it does not explicitly include galaxy size on both axes of the correlation. The $L - R_{\rm e}$ relation has recently become quite popular to study galaxy evolution as a function of redshift (e.g. van der Wel *et al.* 2014), given that it does not involve any kinematic determination and for this reason is much more "economical" to observe than the Faber-Jackson.

Thanks to larger systematic surveys of ETGs it was later discovered that the Faber-Jackson and the Kormendy relations are just two special projections of a plane described by galaxies in $(\log L, \log \sigma, \log R_e)$ coordinates (Dressler *et al.* 1987; Djorgovski & Davis 1987). This plane was aptly named the Fundamental Plane (FP). It was found to hold for all ETGs, including S0s and E galaxies, with a scatter smaller than 20% in R_e (e.g. Jorgensen, Franx & Kjaergaard 1996).

The existence of the FP was interpreted as due to the fact that galaxies satisfy virial equilibrium $M \propto \sigma^2 R_e$, with M the galaxy mass (Faber *et al.* 1987). However the exponents of the FP were found to deviate significantly from the virial predictions, a result confirmed by all numerous subsequent studies (e.g. Hudson *et al.* 1997; Scodeggio *et al.* 1998; Pahre, Djorgovski & de Carvalho 1998; Colless *et al.* 2001; Bernardi *et al.* 2003). In particular, a recent determination of the plane, for the ATLAS^{3D} volume-limited survey (Cappellari *et al.* 2011), with σ_e measured from integral-field stellar kinematics within R_e, gives $L \propto \sigma_e^{1.25} R_e^{0.96}$ (fig. 12 in Cappellari *et al.* 2013b). This deviation of the FP from the virial predictions is called the "tilt" of the FP.

The original FP discovery papers suggested the variation in the stellar M/L as a likely explanation for the tilt. In a brilliant proceedings paper, the Seven Samurai team went as far as stating that "two-dimensionality implies that the virial theorem is the only tight constrain on E structure. (...) implies $(M/L)_e \sim L^{0.24\pm0.04} I_e^{0.00\pm0.06}$. Core and global M/Ls agree well, implying that ellipticals are mainly baryon dominated within R_e and that M/Ls are stellar" (Faber *et al.* 1987). As I will describe in what follows, it took a few decades for all these early insights to be convincingly confirmed.

3. Candidates for the tilt of the Fundamental Plane

The variation of the M/L of the stellar population was immediately recognized as a potential source for the tilt of the FP. This is because systematic changes in the galaxy population were already known, with galaxies becoming older and more metal rich with increasing mass or σ (e.g. Thomas *et al.* 2005, Fig. 2 top-left). This variation can potentially explain a major part of the FP tilt and scatter, predicting larger M/L as a function of σ by a factor of a few, as observed (depending on the photometric band) over the full range of galaxy masses (Prugniel & Simien 1996; Forbes, Ponman & Brown 1998).

The surface brightness profiles of ETGs also display systematic variations as a function of their luminosity. The profiles become more concentrated, or have larger Sersic (1968) indices, for increasing galaxy luminosity (Caon, Capaccioli & D'Onofrio 1993, Graham & Guzmán 2003, Kormendy *et al.* 2009, bottom-left in Fig. 2). At fixed mass, a steeper profile implies a larger σ within the central regions (Ciotti 1991) where the kinematics is observed (typically within a fraction of R_e). The amount of σ variation is again in principle sufficient to explain a major part of the FP tilt (Ciotti, Lanzoni & Renzini 1996; Graham & Colless 1997; Prugniel & Simien 1997; Bertin, Ciotti & Del Principe 2002; Trujillo, Burkert & Bell 2004).

A third potential cause for the FP tilt is the fraction of dark matter within the region where kinematics is observed. The dark matter fraction is expected to increase systematically with mass, for the range of interest of FP studies (e.g. Moster *et al.* 2010, Fig. 2 right panel). This can cause variations in the observed total M/L of an amount again sufficient to produce a significant fraction of the measured tilt (Renzini & Ciotti 1993; Borriello, Salucci & Danese 2003; Tortora *et al.* 2012).

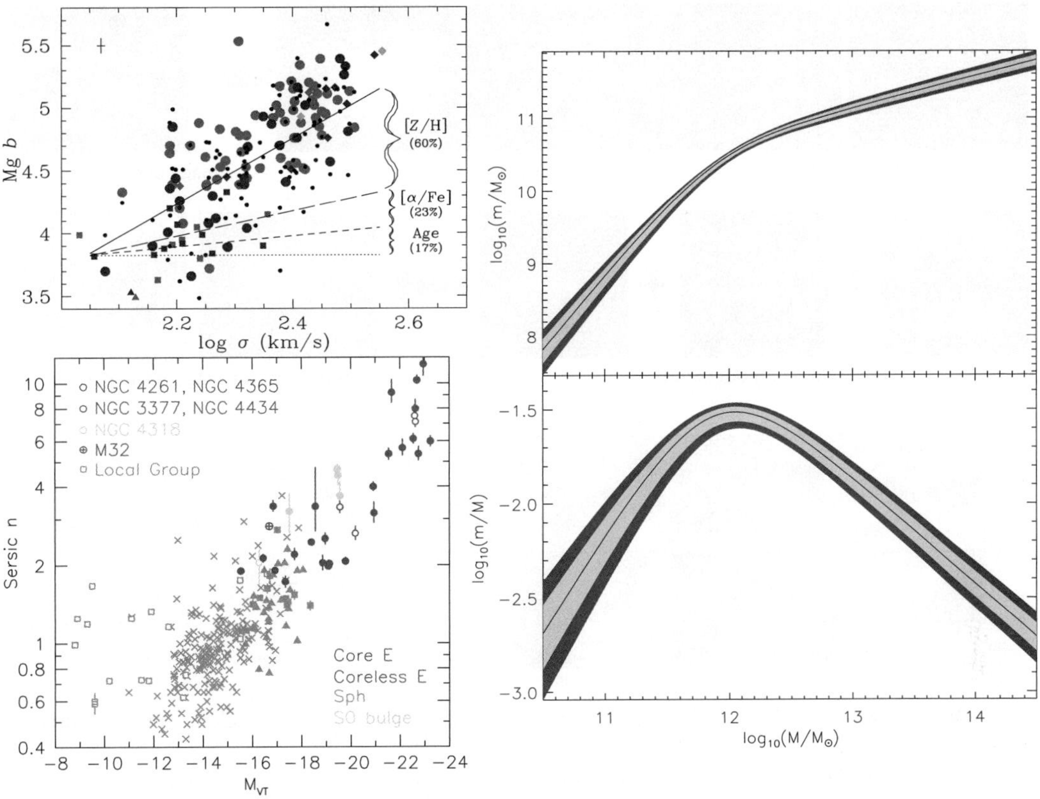

Figure 2. Candidates for the FP tilt. *Top-Left Panel:* Systematic variations in the galaxies stellar population (from Thomas *et al.* 2005). *Bottom-Left Panel:* trends in the non-homology of the surface-brightness profile, as parametrized by the Sersic index (from Kormendy *et al.* 2009). *Right Panel:* Variations in the dark matter fraction (from Moster *et al.* 2010).

4. Measuring unbiased M/L in galaxies

A way to remove most of the uncertainties in the debate about the source of the FP tilt of ETGs consists of accurately modelling stellar population, non-homology and dark matter in a quantitative way. This can be done using dynamical models of the stellar kinematics. Three main techniques have been used in the past decades: (i) Schwarzschild (1979) numerical orbit-superposition technique; (ii) Syer & Tremaine (1996) made-to-measure particle-based approach and (iii) Jeans (1922) hydrodynamic equations.

The most popular has been Schwarzschild (1979) numerical orbit-superposition technique (e.g. Richstone & Tremaine 1988; Rix *et al.* 1997; van der Marel *et al.* 1998; Gebhardt *et al.* 2003; Thomas *et al.* 2004; Cappellari *et al.* 2006; van den Bosch *et al.* 2008), which is able to find the general linear combination of model orbits which best fits the galaxy image and kinematics in exquisite detail (Fig. 3). A downside of the generality of the method is its lack of predictive power. As an example, one can fit axisymmetric Schwarzschild's models to a simulated edge-on barred galaxy. Even in this case, both the galaxy image and the kinematics will be reproduced in great detail. The good fit gives no indication that a bar is present and that the mass model and recovered orbital distribution are significantly in error. Systematic problems in the data can also be easily fitted without raising any concern. Similar strength and limitations are shared by the made-to-measure method (de Lorenzi *et al.* 2007; Dehnen 2009; Long & Mao 2010).

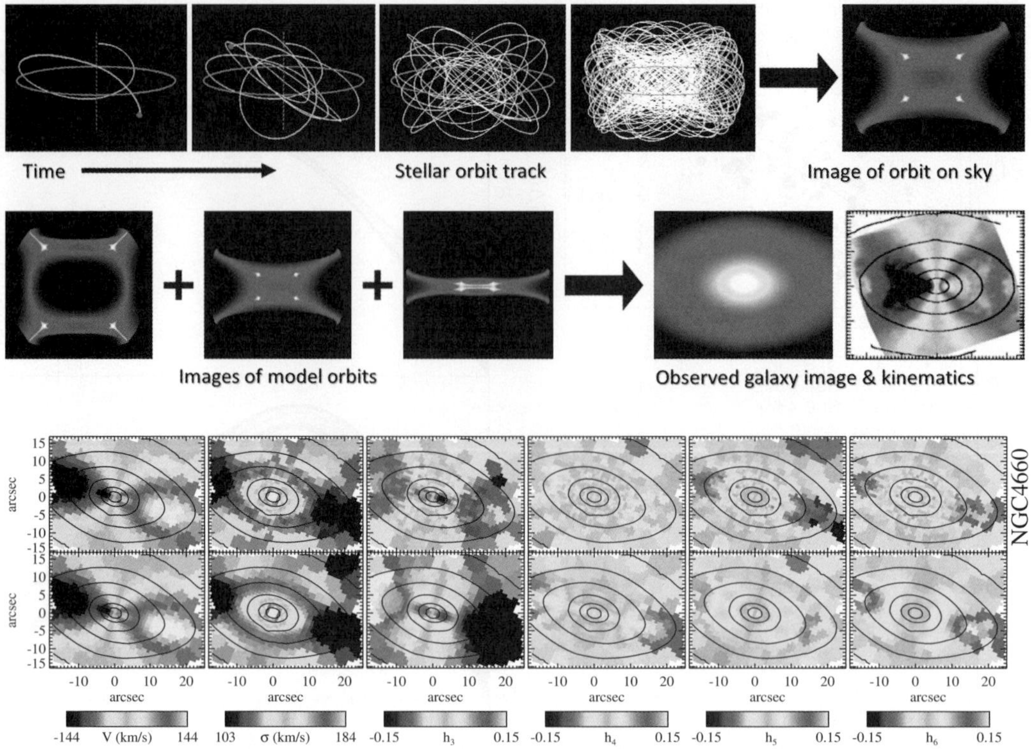

Figure 3. Schwarzschild's orbit-superposition method. *Top Row:* numerical integration of a single orbit in the adopted gravitational potential. After a sufficiently long time the density (of regular orbits) converges to a fixed distribution. *Middle Row:* the method finds the linear combination of thousands of orbits (three representative are shown here) which best fits the galaxy image and stellar kinematics. *Bottom two rows:* data (top) versus model (bottom) comparison. The model can fit the full stellar line-of-sight velocity distribution, here parametrized by the first six Gauss-Hermite moments (from Cappellari *et al.* 2007).

An alternative approach consists of solving the Jeans (1922) equations of stellar hydrodynamics. Contrary to orbit or particle-based methods, with Jeans's method one has to make an assumption about the shape of the velocity ellipsoid in galaxies. Earlier papers assumed a semi-isotropic velocity ellipsoid ($\sigma_z = \sigma_R$), like in Jean's original work (e.g Satoh 1980; Binney, Davies & Illingworth 1990; van der Marel, Binney & Davies 1990; Emsellem, Monnet & Bacon 1994; Magorrian *et al.* 1998). But more recently, Schwarzschild's models based on multi-slit observations (e.g. Cappellari *et al.* 2002; Gebhardt *et al.* 2003; Thomas *et al.* 2009) and especially the constraining power of two-dimensional kinematics (Verolme *et al.* 2002; Krajnović *et al.* 2005; Cappellari *et al.* 2007), have strongly excluded a semi-isotropic form for the velocity ellipsoid in ETGs, making these two-integral models outdated.

However, the availability of good quality two-dimensional stellar kinematics for many galaxies (Emsellem *et al.* 2004; Cappellari *et al.* 2011) also revealed that galaxies have a relatively simple and predictable dynamics within $\sim 1R_{\rm e}$. In fact, by making a simple generalization to Jeans's approach, removing the semi-isotropic assumption by allowing for $\sigma_z \neq \sigma_R$ (Cappellari 2008), one can describe, or essentially "predict", the stellar kinematics of the large majority of real galaxies in surprisingly good detail, using just a couple of free parameters (Scott *et al.* 2009; Cappellari *et al.* 2013b). An added advantage of this three-integral anisotropic Jeans approach is that it is orders of magnitude faster

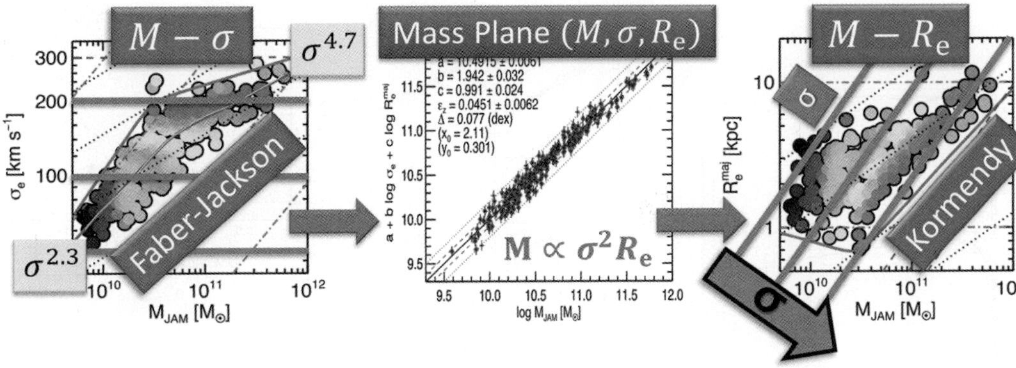

Figure 4. Understanding scaling relations. Replacing luminosity with dynamical mass, the Faber-Jackson shows a clean break at around $M \approx 3 \times 10^{10}\ M_\odot$ (*Left Panel* from Cappellari *et al.* 2013a). The FP becomes the Mass Plane, which follows the virial equation with high accuracy (*Middle Panel* from Cappellari *et al.* 2013b). This implies that the (M, σ) and the (M, R_e) projections contain the same information, apart from a change of coordinates. In both projections, galaxy properties best follow lines of constant σ (*Right Panel*). In the left and right panels, the rainbow colours indicate the M/L.

than the others. It allows for a quick exploration of parameters space for large samples of galaxies using a Bayesian statistical approach.

5. Understanding scaling relations

Employing the semi-isotropic Jeans approach, and stellar kinematics of 37 galaxies from major/minor axis long-slit observations, it was found that the M/L trend with galaxy mass remains nearly unchanged when one includes the effects of galaxy non-homology (van der Marel 1991; Magorrian *et al.* 1998). The model accuracy was improved by fitting models to two-dimensional stellar kinematics. Using a sample of 25 galaxies and both the Schwarzschild and Jeans approaches, Cappellari *et al.* (2006) found that the $(M/L) - \sigma$ relation is extremely tight and accounts for the entire scatter and tilt of the FP. In other words, when replacing luminosity with mass in the FP, the coefficients of the derived Mass Plane matched the virial predictions $M \propto \sigma^2 R_e$ within the errors. Independent confirmations were found by strong lensing studies (Bolton *et al.* 2007; Auger *et al.* 2010a). This result was strengthened by modelling the 260 ETGs of the ATLAS^{3D} sample with two-dimensional stellar kinematics. Galaxies were found to follow the virial predictions with high accuracy (Fig. 4 middle). However, the study also pointed out the significant dependence of the plane coefficients on the technique used to measure them (Cappellari *et al.* 2013b). This sensitivity can explain the apparent contrast between some of the past studies of the FP tilt.

Unbiased studies using detailed dynamical models found that the dynamically-derived $(M_*/L)_{\rm dyn}$ was related to the $(M_*/L)_{\rm pop}$ inferred from stellar population models (Gerhard *et al.* 2001). This confirmed that at least part of the FP tilt is due to stellar population variations. It agrees with the fact that the scatter around the FP is also linked to variations in the stellar population (Graves, Faber & Schiavon 2009; Falcón-Barroso *et al.* 2011; Springob *et al.* 2012; Magoulas *et al.* 2012).

However, even improving the accuracy of the models using two-dimensional kinematics, the relation between dynamically-derived *total* M/L and the stellar population $(M_*/L)_{\rm pop}$ still showed significant systematic deviations (Cappellari *et al.* 2006). These could only be explained by either dark matter or IMF variations between galaxies.

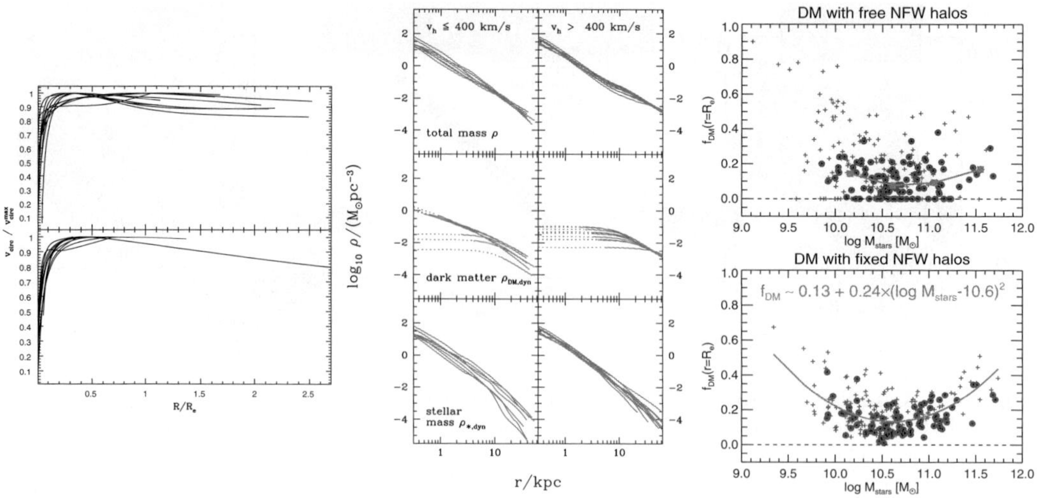

Figure 5. Dark matter from stellar dynamical models. *Left Panels:* The circular velocity profiles of ETGs, as inferred from the models, are nearly flat within $\sim 2R_e$ (from Gerhard *et al.* 2001). *Middle Panels:* The total mass profiles are approximately isothermal ($\rho \propto r^{-2}$) within $\sim 2R_e$ (from Thomas *et al.* 2011). *Right Panels:* Either leaving the halo normalization as a free parameter in the models (top), or constraining it to the value predicted by ΛCDM, the inferred dark matter fractions within R_e must be small, to fit the kinematics (from Cappellari *et al.* 2013b).

Deviations between accurate determinations of the stellar and total masses were also found using independent strong lensing techniques (e.g. Auger *et al.* 2010a).

To discriminate between dark matter and/or the IMF as the reason for the observed mass discrepancies, one needs to construct dynamical models which explicitly account for both the luminous and dark matter contributions. Unfortunately the problem is intrinsically quite degenerate. Still, using long-slit data and general models for two samples of about 20 galaxies, different studies appeared to agree that (i) dark matter represents a minor fraction of the total, within a sphere of radius $r \sim R_e$; (ii) the *total* mass profile is nearly isothermal ($\rho \propto r^{-2}$) and produces almost flat circular rotation curves (Fig. 5 left and middle), similarly to spiral galaxies. This was measured out to the median radius $r \sim 2R_e$ sampled by the kinematics (Gerhard *et al.* 2001; Thomas *et al.* 2011) and in a few cases out to larger radii (e.g. Morganti *et al.* 2013). The dark matter content within R_e was better quantified with the modelling of the 260 ETGs of the ATLAS^{3D} sample. A median dark matter fraction $f_{DM}(r = R_e)$ as low as 13% was measured for the full sample (Fig. 5 right).

Thanks to the large sample and two-dimensional stellar kinematics, the ATLAS^{3D} study was able to show that the systematic trend in the discrepancy between $(M_*/L)_{dyn}$ and $(M_*/L)_{pop}$ could not be explained by a variation in the dark matter fraction. The most likely reason was then a systematic variation of the stellar initial mass function (IMF). This was inferred to vary in mass normalization from Milky-Way type (Kroupa 2001; Chabrier 2003) to heavier than Salpeter (1955) type, over the full mass range (Cappellari *et al.* 2012). This systematic trend was consistent with indications of a "heavy" IMF in massive ETGs from either stellar population (van Dokkum & Conroy 2010) or strong gravitational lensing (Auger *et al.* 2010b). This variation seems to be naturally explained by some theoretical models (e.g. Chabrier, Hennebelle & Charlot 2014).

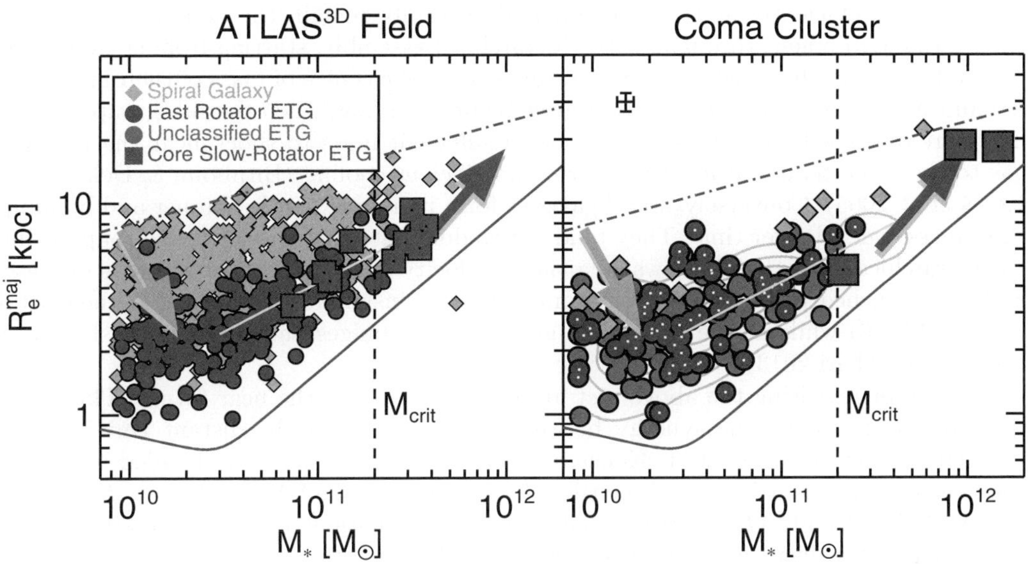

Figure 6. The bulid-up of the Mass-Size relation. The *Left Panel* shows a field sample. The *Right Panel* shows an identically-selected sample in one of the densest environments in the Universe. The magenta arrow qualitatively indicates the evolutionary track due to bulge growth and environmental quenching. In the dense environment, spirals are replaced by fast rotator ETGs, which have the same mass and size distribution as in the field sample. The red arrow shows the dry merging and halo quenching track, for slow rotators. These build up larger masses in denser environments (from Cappellari 2013).

6. Implications for galaxy formation

The fact that the FP is due to virial equilibrium implies that only a modest amount of information on galaxy formation is provided by the mere existence of the plane. Instead, most constraints on galaxy formation are encoded in the distribution of galaxy properties *within* the mass plane. A key empirical finding is that, within the plane, nearly all variation in galaxy properties is best described by a trend with the galaxy σ, rather than other global parameters like dynamical mass, size, surface brightness or (Sersic 1968) index (Fig. 4). This is true for the M/L (Cappellari *et al.* 2006), stellar population indicators (Graves, Faber & Schiavon 2009; Poggianti *et al.* 2013) as well as for the molecular gas fraction, colour and IMF (Cappellari *et al.* 2013a).

It has become clear that σ is a simple empirical tracer of the galaxy bulge *mass* fraction (Cappellari *et al.* 2013a). And the correlations of galaxy properties with σ describe a link between the mass growth in the central bulge and the cessation of the galaxy star formation. Similar results and conclusions are reached when σ is replaced by a density measure within a fixed aperture (Cheung *et al.* 2012; Fang *et al.* 2013). This trend between bulge and quenching persists over the full range of galaxies morphological types, going smoothly from nearly bulge-less spiral galaxies, to the most dense disky-ellipticals fast-rotator ETGs (Fig. 6). However the bulge trend only exists below a characteristic mass $M_{\rm crit} \approx 2 \times 10^{11}$ $M_\odot$. Above $M_{\rm crit}$ a different process is at work, which produces massive slow rotator ETGs with cores in their surface brightness (Fig. 6), preferentially lying near the centre of clusters or infalling groups (Cappellari 2013; Fogarty *et al.* 2014).

The distinction between the first, bulge-related, quenching and the second, halo-related quenching, can be understood within the framework of hierarchical morphological evolution (De Lucia *et al.* 2012), where the massive slow rotating galaxies tend to form most of their mass efficiently at high redshift and generally remain the dominant galaxy of their own environment throughout the hierarchical assembly, starting from small groups to build more massive clusters. Slow rotators sink to the centre of mass of their groups by dynamical friction, where they further grow by dry merging, when the sub-groups merge. They are quenched when the halo reaches a sufficient mass to shock heat the infalling gas to the virial temperature, which prevents efficient cooling (Birnboim & Dekel 2003; Kereš *et al.* 2005). Conversely, spiral galaxies form in a more gentle manner and build up their mass over a longer time. They tend not to dominate the mass of their group during the hierarchical growth. They swarm at large velocity around the cluster centre of mass, and are quenched by the cluster environment, with insignificant increase in their total masses (Fig. 6) while at the same time growing their bulges and becoming fast rotator ETGs (Cappellari 2013).

Defining or revisiting the above picture, is a challenge for the near future. This effort is currently driven by the synergy between large surveys of the distant Universe like CANDLES (Grogin *et al.* 2011; Koekemoer *et al.* 2011) and massive detailed studies of the nearby Universe using two-dimensional spectroscopy like SAMI (Croom *et al.* 2012) and MaNGA (Bundy *et al.* 2015), combined with simulations of ever increasing spatial resolution and realism (e.g. Vogelsberger *et al.* 2014).

Acknowledgements

I acknowledge support from a Royal Society University Research Fellowship.

References

Auger, M. W., Treu, T., Bolton, A. S., Gavazzi, R., Koopmans, L. V. E., Marshall, P. J., Moustakas, L. A., & Burles, S., 2010a, *ApJ*, 724, 511
Auger, M. W., Treu, T., Gavazzi, R., Bolton, A. S., Koopmans, L. V. E., & Marshall, P. J., 2010b, *ApJ*, 721, L163
Bernardi, M. *et al.*, 2003, *AJ*, 125, 1866
Bertin, G., Ciotti, L., & Del Principe, M., 2002, *A&A*, 386, 149
Binney, J. J., Davies, R. L., & Illingworth, G. D., 1990, *ApJ*, 361, 78
Birnboim, Y. & Dekel, A., 2003, *MNRAS*, 345, 349
Bolton, A. S., Burles, S., Treu, T., Koopmans, L. V. E., & Moustakas, L. A., 2007, *ApJ*, 665, L105
Borriello, A., Salucci, P., & Danese, L., 2003, *MNRAS*, 341, 1109
Boylan-Kolchin, M., Ma, C.-P., & Quataert, E., 2006, *MNRAS*, 369, 1081
Bundy, K. *et al.*, 2015, *ApJ*, 798, 7
Caon, N., Capaccioli, M., & D'Onofrio, M., 1993, *MNRAS*, 265, 1013
Cappellari, M., 2008, *MNRAS*, 390, 71
Cappellari, M., 2013, *ApJ*, 778, L2
Cappellari, M. *et al.*, 2006, *MNRAS*, 366, 1126
Cappellari, M. *et al.*, 2007, *MNRAS*, 379, 418
Cappellari, M. *et al.*, 2011, *MNRAS*, 413, 813
Cappellari, M. *et al.*, 2012, *Nature*, 484, 485
Cappellari, M. *et al.*, 2013a, *MNRAS*, 432, 1862
Cappellari, M. *et al.*, 2013b, *MNRAS*, 432, 1709
Cappellari, M., Verolme, E. K., van der Marel, R. P., Kleijn, G. A. V., Illingworth, G. D., Franx, M., Carollo, C. M., & de Zeeuw, P. T., 2002, *ApJ*, 578, 787
Chabrier, G., 2003, *PASP*, 115, 763
Chabrier, G., Hennebelle, P., & Charlot, S., 2014, *ApJ*, 796, 75
Cheung, E. *et al.*, 2012, *ApJ*, 760, 131

Ciotti, L., 1991, *A&A*, 249, 99
Ciotti, L., Lanzoni, B., & Renzini, A., 1996, *MNRAS*, 282, 1
Colless, M., Saglia, R. P., Burstein, D., Davies, R. L., McMahan, R. K., & Wegner, G., 2001, *MNRAS*, 321, 277
Courteau, S. *et al.*, 2014, *Reviews of Modern Physics*, 86, 47
Croom, S. M. *et al.*, 2012, *MNRAS*, 421, 872
de Lorenzi, F., Debattista, V. P., Gerhard, O., & Sambhus, N., 2007, *MNRAS*, 376, 71
De Lucia, G., Weinmann, S., Poggianti, B. M., Aragón-Salamanca, A., & Zaritsky, D., 2012, *MNRAS*, 423, 1277
Dehnen, W., 2009, *MNRAS*, 395, 1079
Djorgovski, S. & Davis, M., 1987, *ApJ*, 313, 59
Dressler, A., Lynden-Bell, D., Burstein, D., Davies, R. L., Faber, S. M., Terlevich, R., & Wegner, G., 1987, *ApJ*, 313, 42
Emsellem, E. *et al.*, 2004, *MNRAS*, 352, 721
Emsellem, E., Monnet, G., & Bacon, R., 1994, *A&A*, 285, 723
Faber, S. M., Dressler, A., Davies, R. L., Burstein, D., & Lynden-Bell, D., 1987, in *Nearly Normal Galaxies. From the Planck Time to the Present*, Faber S. M., ed., Springer-Verlag, New York, pp. 175–183
Faber, S. M. & Jackson, R. E., 1976, *ApJ*, 204, 668
Falcón-Barroso, J. *et al.*, 2011, *MNRAS*, 417, 1787
Fang, J. J., Faber, S. M., Koo, D. C., & Dekel, A., 2013, *ApJ* in press
Fogarty, L. M. R. *et al.*, 2014, *MNRAS*, 443, 485
Forbes, D. A., Ponman, T. J., & Brown, R. J. N., 1998, *ApJ*, 508, L43
Gebhardt, K. *et al.*, 2003, *ApJ*, 583, 92
Gerhard, O., Kronawitter, A., Saglia, R. P., & Bender, R., 2001, *AJ*, 121, 1936
Graham, A. & Colless, M., 1997, *MNRAS*, 287, 221
Graham, A. W. & Guzmán, R., 2003, *AJ*, 125, 2936
Graves, G. J., Faber, S. M., & Schiavon, R. P., 2009, *ApJ*, 698, 1590
Grogin, N. A. *et al.*, 2011, *ApJS*, 197, 35
Hopkins, P. F., Bundy, K., Hernquist, L., Wuyts, S., & Cox, T. J., 2010, *MNRAS*, 401, 1099
Hudson, M. J., Lucey, J. R., Smith, R. J., & Steel, J., 1997, *MNRAS*, 291, 488
Jeans, J. H., 1922, *MNRAS*, 82, 122
Jorgensen, I., Franx, M., & Kjaergaard, P., 1996, *MNRAS*, 280, 167
Kereš, D., Katz, N., Weinberg, D. H., & Davé, R., 2005, *MNRAS*, 363, 2
Koekemoer, A. M. *et al.*, 2011, *ApJS*, 197, 36
Kormendy, J., 1977, *ApJ*, 218, 333
Kormendy, J., Fisher, D. B., Cornell, M. E., & Bender, R., 2009, *ApJS*, 182, 216
Krajnović, D., Cappellari, M., Emsellem, E., McDermid, R. M., & de Zeeuw, P. T., 2005, *MNRAS*, 357, 1113
Kroupa, P., 2001, *MNRAS*, 322, 231
Long, R. J. & Mao, S., 2010, *MNRAS*, 405, 301
Magorrian, J. *et al.*, 1998, *AJ*, 115, 2285
Magoulas, C. *et al.*, 2012, *MNRAS*, 427, 245
Morganti, L., Gerhard, O., Coccato, L., Martinez-Valpuesta, I., & Arnaboldi, M., 2013, *MNRAS*, 431, 3570
Moster, B. P., Somerville, R. S., Maulbetsch, C., van den Bosch, F. C., Macciò, A. V., Naab, T., & Oser, L., 2010, *ApJ*, 710, 903
Naab, T., Johansson, P. H., & Ostriker, J. P., 2009, *ApJ*, 699, L178
Oser, L., Naab, T., Ostriker, J. P., & Johansson, P. H., 2012, *ApJ*, 744, 63
Pahre, M. A., Djorgovski, S. G., & de Carvalho, R. R., 1998, *AJ*, 116, 1591
Planck Collaboration, 2014, *A&A*, 571, A16
Poggianti, B. M. *et al.*, 2013, *ApJ*, 762, 77
Porter, L. A., Somerville, R. S., Primack, J. R., & Johansson, P. H., 2014, *MNRAS*, 444, 942
Prugniel, P. & Simien, F., 1996, *A&A*, 309, 749
Prugniel, P. & Simien, F., 1997, *A&A*, 321, 111

Renzini, A. & Ciotti, L., 1993, *ApJ*, 416, L49
Richstone, D. O. & Tremaine, S., 1988, *ApJ*, 327, 82
Rix, H.-W., de Zeeuw, P. T., Cretton, N., van der Marel, R. P., & Carollo, C. M., 1997, *ApJ*, 488, 702
Robertson, B., Bullock, J. S., Cox, T. J., Di Matteo, T., Hernquist, L., Springel, V., & Yoshida, N., 2006, *ApJ*, 645, 986
Salpeter, E. E., 1955, *ApJ*, 121, 161
Satoh, C., 1980, *PASJ*, 32, 41
Schwarzschild, M., 1979, *ApJ*, 232, 236
Scodeggio, M., Gavazzi, G., Belsole, E., Pierini, D., & Boselli, A., 1998, *MNRAS*, 301, 1001
Scott, N. *et al.*, 2009, *MNRAS*, 398, 1835
Sersic, J. L., 1968, *Atlas de galaxias australes.* Cordoba, Argentina: Observatorio Astronomico, 1968
Springob, C. M. *et al.*, 2012, *MNRAS*, 420, 2773
Syer, D. & Tremaine, S., 1996, *MNRAS*, 282, 223
Thomas, D., Maraston, C., Bender, R., & Mendes de Oliveira, C., 2005, *ApJ*, 621, 673
Thomas, J. *et al.*, 2009, *MNRAS*, 393, 641
Thomas, J. *et al.*, 2011, *MNRAS*, 415, 545
Thomas, J., Saglia, R. P., Bender, R., Thomas, D., Gebhardt, K., Magorrian, J., & Richstone, D., 2004, *MNRAS*, 353, 391
Tonry, J. L., Dressler, A., Blakeslee, J. P., Ajhar, E. A., Fletcher, A. B., Luppino, G. A., Metzger, M. R., & Moore, C. B., 2001, *ApJ*, 546, 681
Tortora, C., La Barbera, F., Napolitano, N. R., de Carvalho, R. R., & Romanowsky, A. J., 2012, *MNRAS*, 425, 577
Trujillo, I., Burkert, A., & Bell, E. F., 2004, *ApJ*, 600, L39
van den Bosch, R. C. E., van de Ven, G., Verolme, E. K., Cappellari, M., & de Zeeuw, P. T., 2008, *MNRAS*, 385, 647
van der Marel, R. P., 1991, *MNRAS*, 253, 710
van der Marel, R. P., Binney, J., & Davies, R. L., 1990, *MNRAS*, 245, 582
van der Marel, R. P., Cretton, N., de Zeeuw, P. T., & Rix, H.-W., 1998, *ApJ*, 493, 613
van der Wel, A. *et al.*, 2014, *ApJ*, 788, 28
van Dokkum, P. G. & Conroy, C., 2010, *Nature*, 468, 940
Verolme, E. K. *et al.*, 2002, *MNRAS*, 335, 517
Vogelsberger, M. *et al.*, 2014, *Nature*, 509, 177

Galaxy Masses as Constraints of Formation Models
Proceedings IAU Symposium No. 311, 2014
M. Cappellari & S. Courteau, eds.

doi:10.1017/S1743921315003336

The Outer Halos of Early-Type Galaxies

Ortwin Gerhard[1], Magda Arnaboldi[2] and Alessia Longobardi[1]

[1]Max-Planck-Institut für extraterrestrische Physik, Postfach 1312, Giessenbachstr. 1, 85741 Garching, Germany. email: gerhard@mpe.mpg.de, alongobardi@mpe.mpg.de
[2]European Southern Observatory, Karl-Schwarzschild-Strasse 2, D-85748 Garching, Germany. e-mail: marnabol@eso.org

Abstract. The outer halos of massive early-type galaxies (ETGs) are dark matter dominated and may have formed by accretion of smaller systems during galaxy evolution. Here a brief report is given of some recent work on the kinematics, angular momentum, and mass distributions of simulated ETG halos, and of corresponding properties of observed halos measured with planetary nebulae (PNs) as tracers. In the outermost regions of the Virgo-central galaxy M87, the PN data show that the stellar halo and the co-spatial intracluster light are distinct kinematic components.

---

1. Stellar and dark matter halos in simulated ETGs

Wu *et al.* (2014) studied the luminous and dark matter halos in 42 simulated early-type galaxies (ETGs) with stellar masses from $2.0 \times 10^{10} M_\odot\ h^{-1}$ to $3.4 \times 10^{11} M_\odot\ h^{-1}$. These simulated galaxies were obtained from N-body/hydrodynamical zoom re-simulations of selected halos from a dark matter cosmological simulation, including a model for star formation and feedback (Oser *et al.* 2010, 2012).

The dark matter halo density profiles of the simulated ETGs follow simple power-law models, with flat dark matter circular velocity curves (CVCs) for lower mass systems and rising CVCs for high-mass halos, corresponding to slightly falling to flat total CVCs respectively. The dark matter fractions of these systems are in the range 15-30% at the stellar half-mass radius R_e and increase to 40-65% at $5\,R_e$. The short axes of simulated galaxies and their host dark matter halos are well-aligned and their (short-to-long) axis ratios are correlated.

Line-of-sight (LOS) velocity fields obtained with a time-smoothing technique out to large radii (see examples in Fig. 1) show that the rotation properties at small and large radii are correlated. Radial profiles for the cumulative specific angular momentum parameter $\lambda(R)$ are nearly flat or slightly rising, with values in [0.06, 0.75] from $2\,R_e$ to $5\,R_e$. A few cases show local maxima in $|\bar{v}|/\sigma(R)$.

Many of the properties of these simulated ETGs agree remarkably well with observations of ETGs at large radii, such as briefly summarized in the next section.

2. Kinematics, angular momentum, and dark matter mass in ETG halos from PNs

Planetary nebulae (PNs) are excellent tracers for the kinematics in the faint outskirts of ETGs where kinematic measurements from integrated light are no longer possible. Various techniques have been implemented to measure radial velocities from their bright [OIII]λ5007Å line (e.g., Hui *et al.* 1995; Arnaboldi *et al.* 1998; Méndez *et al.* 2001; McNeil *et al.* 2010). With the specially built Planetary Spectrograph (PN.S) instrument (Douglas *et al.* 2002), a survey of nearby ($D < 20$ Mpc) early-type galaxy halos in the Northern hemisphere has now been completed. The new sample expands on earlier results based

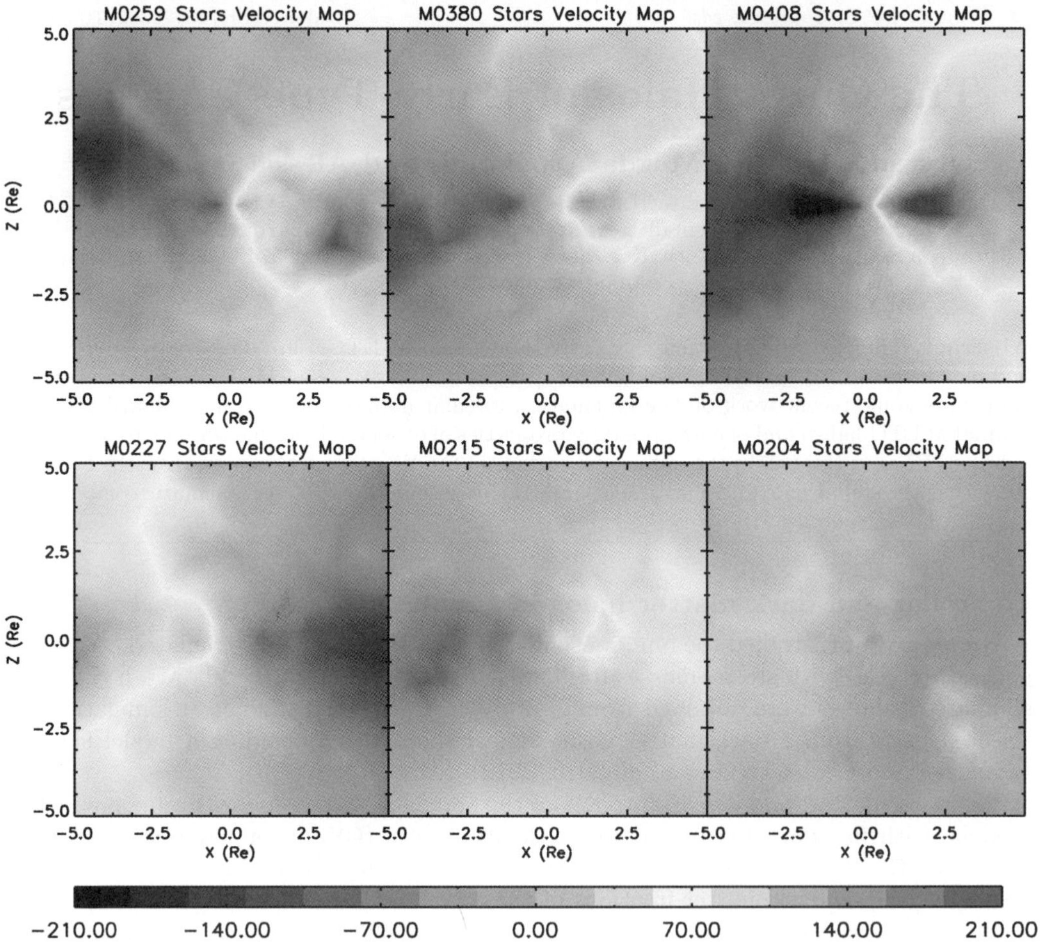

Figure 1. Edge-on mean LOS velocity maps for the stellar components of 6 selected model galaxies. The upper panels show three simulated galaxies with large scale rotation; the galaxy in the top right panel has a strong disk component from a gas rich major merger. The lower panels show three simulated galaxies with a dissipationless merging history. The galaxy in the bottom right panel shows weak major and minor axis rotation; it has a formation history with many minor mergers. From Wu *et al.* (2014).

on 6 lenticulars and 13 ellipticals (Coccato *et al.* 2009; Cortesi *et al.* 2013), to 11 S0s and 30 ellipticals with extended kinematics measured from 2-10R_e.

From the PN data one can measure the specific angular momentum out to these large radii (Fig. 2, from Cortesi *et al.* 2013) and the frequency of angular momentum misalignment between the inner region and the outer halo. One interesting, still not fully understood result is the apparent dichotomy between galaxies with quasi-Keplerian, steeply falling, velocity dispersion profiles and galaxies with nearly flat velocity dispersion profiles (Coccato *et al.* 2009). Dynamical modeling of a few of the quasi-Keplerian systems with the NMAGIC made-to-measure method (de Lorenzi *et al.* 2007) has shown a strong mass-anisotropy degeneracy (de Lorenzi *et al.* 2008, 2009), with the range of best dark matter halo profiles (Fig. 3; Morganti *et al.* 2013) consistent with those of the lower mass, simulated ETGs of Wu *et al.* (2014); see also Deason *et al.* (2012). By contrast, for massive ETGs with nearly flat dispersion profiles the inferred total CVCs are nearly

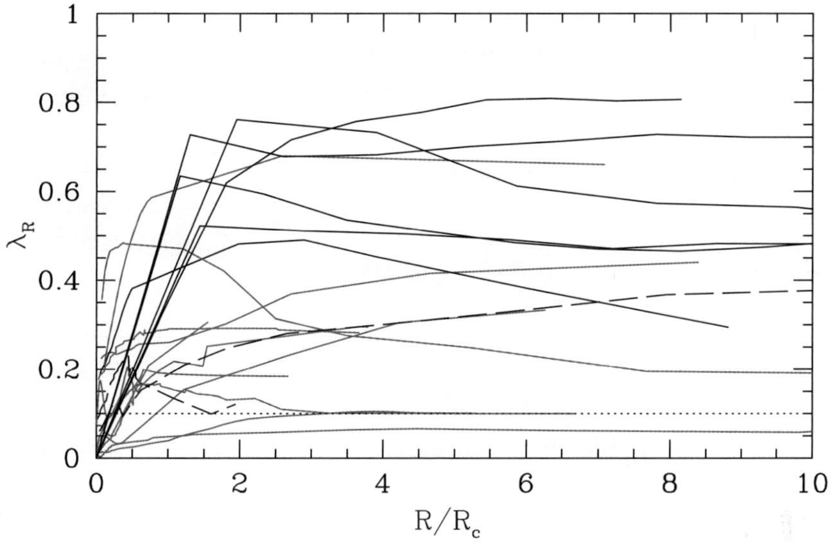

Figure 2. The λ_R parameter measuring enclosed specific angular momentum as a function of radius, for a sample of early-type galaxies. Black solid lines are S0s from Cortesi *et al.* (2013), black dashed lines are S0s from Coccato *et al.* (2009), and red lines are ellipticals from Coccato *et al.* (2009). The dotted line shows the suggested separation between slow and fast rotators (Emsellem *et al.* 2007). From Cortesi *et al.* (2013).

constant with radius (e.g., Das *et al.* 2011). Full analysis of the PN.S survey data is still on-going.

3. Distinct stellar halo and intracluster light components in the outskirts of the giant Virgo ETG M87

Brightest cluster galaxies (BCGs) have extended halos often embedded in diffuse intracluster light (ICL; e.g., Gonzalez *et al.* 2005). The ICL consists of stars moving in the cluster potential, whereas halo stars would be defined as bound to the central galaxy. However, if the BCG is located in the cluster center, it is not clear whether a clear distinction between halo and ICL is generally possible (Dolag *et al.* 2010; Cooper *et al.* 2013). An interesting case is NGC 3311 in the Hydra I cluster, where the very rapid transition of the velocity dispersion profile to cluster values suggests a continuous transition from halo to ICL (Ventimiglia *et al.* 2010).

Doherty *et al.* (2009) first investigated the transition from halo to ICL in the outermost regions of the Virgo cluster. With a small sample of PNs, they found no halo PNs beyond 150 kpc radius, and a rapidly decreasing velocity dispersion profile towards this boundary, while ICL PNs with large velocity dispersion where seen at similar radii. Based on a large sample of PN candidates indentified through on/off-band imaging (Longobardi *et al.* 2013), and kinematic follow-up with the VLT FLAMES instrument (Longobardi *et al.* 2014), it was possible to separate PNs from the M87 halo and ICL with a robust kinematic technique. Fig. 4 shows that both components coexist at similar radii with different density profiles. The relative abundance of PNs per unit bolometric luminosity and the PN luminosity function furthermore suggest that the outer M87 halo and the ICL in the Virgo core are different stellar populations.

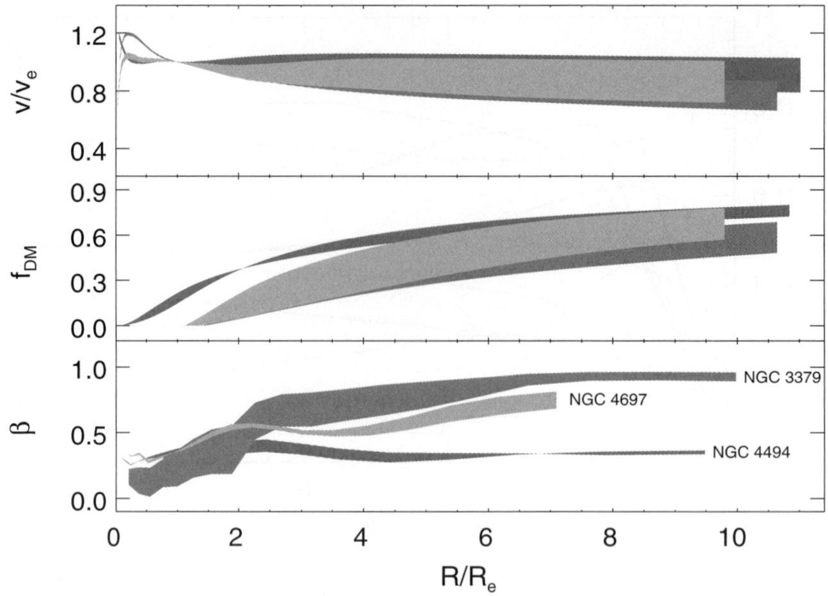

Figure 3. Radial profiles of circular velocity normalized by its value at $1R_e$ (top), dark matter fraction (middle), and anisotropy parameter (bottom), for the range of valid NMAGIC models obtained fitting the data of NGC 4494 (*violet*, Morganti *et al.* 2013), NGC 4697 (*orange*, de Lorenzi *et al.* 2008), and NGC 3379 (*green*, de Lorenzi *et al.* 2009). From Morganti *et al.* (2013).

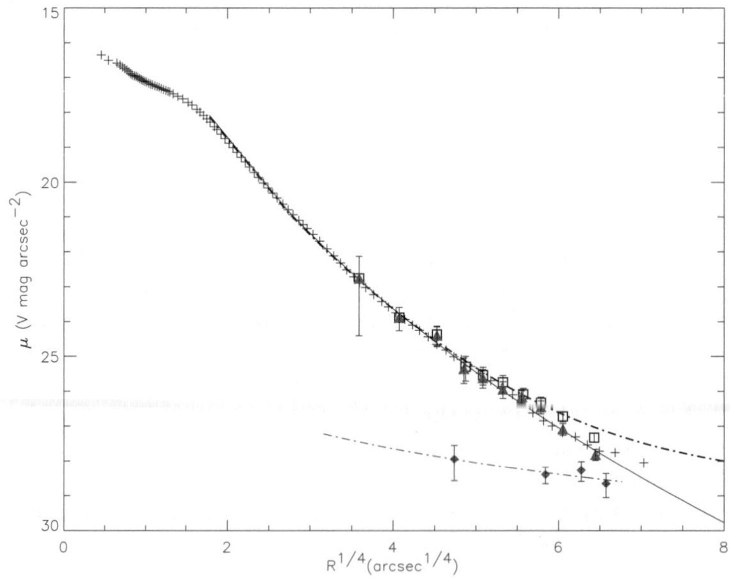

Figure 4. The logarithmic PN number density profile of the M87 halo (red triangles) and ICL PNs (blue diamonds), corrected for incompleteness as a function of the distance from the M87 centre. For comparison, the surface brightness profile from Kormendy *et al.* (2009) is shown as crosses. The ICL density decreases towards larger radii as $I_{ICL} \propto R^{-0.4}$ (dashed blue line). Black squares show the composite halo plus ICL PN number density profile, well-modelled by the two component model (dot-dashed black line, see Longobardi *et al.* (2014) for more details). Error bars account for 1 σ uncertainties from counting statistics. From Longobardi *et al.* (2014).

References

Arnaboldi, M., *et al.* 1998, *ApJ*, 507, 759
Coccato, L., *et al.* 2009, *MNRAS*, 394, 1249
Cooper, A. P., *et al.* 2013, *MNRAS*, 434, 3348
Cortesi, A., *et al.* 2013, *MNRAS*, 432, 1010
Das, P., *et al.* 2011, *MNRAS*, 415, 1244
de Lorenzi, F., Debattista, V. P., Gerhard, O., & Sambhus, N. 2007, *MNRAS*, 376, 71
de Lorenzi, F., *et al.* 2008, *MNRAS*, 385, 1729
—. 2009, *MNRAS*, 395, 76
Deason, A. J., Belokurov, V., Evans, N. W., & McCarthy, I. G. 2012, *ApJ*, 748, 2
Doherty, M., *et al.* 2009, *A&A*, 502, 771
Dolag, K., Murante, G., & Borgani, S. 2010, *MNRAS*, 405, 1544
Douglas, N. G., *et al.* 2002, *PASP*, 114, 1234
Emsellem, E., *et al.* 2007, *MNRAS*, 379, 401
Gonzalez, A. H., Zabludoff, A. I., & Zaritsky, D. 2005, *ApJ*, 618, 195
Hui, X., Ford, H. C., Freeman, K. C., & Dopita, M. A. 1995, *ApJ*, 449, 592
Kormendy, J., Fisher, D. B., Cornell, M. E., & Bender, R. 2009, *ApJS*, 182, 216
Longobardi, A., *et al.* 2013, *A&A*, 558, A42
Longobardi, A., Arnaboldi, M., Gerhard, O., & Hanuschik, R. 2014, *A&A*, submitted
McNeil, E. K., *et al.* 2010, *A&A*, 518, A44
Méndez, R. H., *et al.* 2001, *ApJ*, 563, 135
Morganti, L., *et al.* 2013, *MNRAS*, 431, 3570
Oser, L., Naab, T., Ostriker, J. P., & Johansson, P. H. 2012, *ApJ*, 744, 63
Oser, L., *et al.* 2010, *ApJ*, 725, 2312
Ventimiglia, G., Gerhard, O., Arnaboldi, M., & Coccato, L. 2010, *A&A*, 520, L9+
Wu, X., *et al.* 2014, *MNRAS*, 438, 2701

Galaxy Masses as Constraints of Formation Models
Proceedings IAU Symposium No. 311, 2014
M. Cappellari & S. Courteau, eds.

doi:10.1017/S1743921315003348

Massive Elliptical Galaxies: BH Scouring or a Bottom-Heavy IMF?

Jens Thomas[1,2], Roberto Saglia[1,2], Ralf Bender[1,2], Peter Erwin[1,2] and Maximilian Fabricius[1,2]

[1]Max-Planck-Institut für extraterrestrische Physik, Giessenbachstr. 1, 85748 Garching
[2]Universitäts-Sternwarte München, Scheinerstr. 1, 81679 München

Abstract. We present indirect constraints on the stellar initial-mass-function (IMF) in nine massive elliptical galaxies with $\sigma \approx 300$ km/s, via a comparison of dynamical and stellar-population based stellar masses. We use adaptive-optics assisted, high resolution kinematical data from the SINFONI Search for Supermassive Black Holes that allow us to constrain the dynamical stellar mass-to-light ratio in the very centre of each galaxy. Hence we measure the IMF in a galaxy region where the stellar mass dominates over dark matter, minimising any potential degeneracy between the two mass components. In six of our galaxies – those which have depleted stellar cores – we find an IMF consistent with the one measured in the Milky-Way via direct star counts. The three remaining, power-law galaxies have instead stellar masses about a factor of two times larger than expected from a Milky-Way type IMF, indicating either a more bottom-heavy IMF (like, e.g., the Salpeter IMF) or a dark-matter distribution that is degenerate with the stellar mass down to the very centres of these galaxies. The bottom-light IMF in our core galaxies is surprising in view of previous studies that suggested a systematic IMF variation where early-type galaxies with $\sigma \approx 300$ km/s have a Salpeter or even more dwarf-dominated IMF. Core galaxies are particularly important since their unique central orbital structure offers an independent crosscheck for the dynamical models. Our models with a bottom-light IMF are consistent with the distribution of orbits predicted by SMBH-binary core-formation models. This indicates that spatially well resolved central kinematical data are important for determining unbiased dynamical stellar mass-to-light ratios. Our results imply either that the IMF in massive galaxies varies over a wider range than previously anticipated, and is not the same in core and power-law ellipticals, or else that there are systematic variations in the distribution of dark matter among massive early-type galaxies.

Keywords. galaxies: elliptical and lenticular, cD - galaxies: evolution - galaxies: formation

1. Introduction

Several recent studies have investigated whether the IMF systematically varies with galaxy mass, in particular among elliptical galaxies. These analyses are mostly based on two independent approaches. The first is an indirect method, where galaxy stellar masses are determined from stellar population synthesis models that actually do not resolve the IMF: the IMF is adjusted until the population-synthesis mass-to-light ratio matches independent constraints from dynamics and/or lensing. The second, direct method uses features in galaxy spectra that are particularly sensitive to the presence of, e.g., dwarf stars to determine the IMF directly from spectroscopic observations. Both approaches come to the same result: although there is significant galaxy to galaxy scatter, lower mass early-type galaxies (with dispersions $\sigma \approx 200$ km/s) seem to be roughly consistent with a Milky-Way type IMF (e.g. a Kroupa or Chabrier IMF). In high-dispersion elliptical galaxies, however, stellar mass-to-light ratios are about a factor of 2 times higher than expected from a Kroupa IMF. Direct studies specifically indicate that the IMF in massive

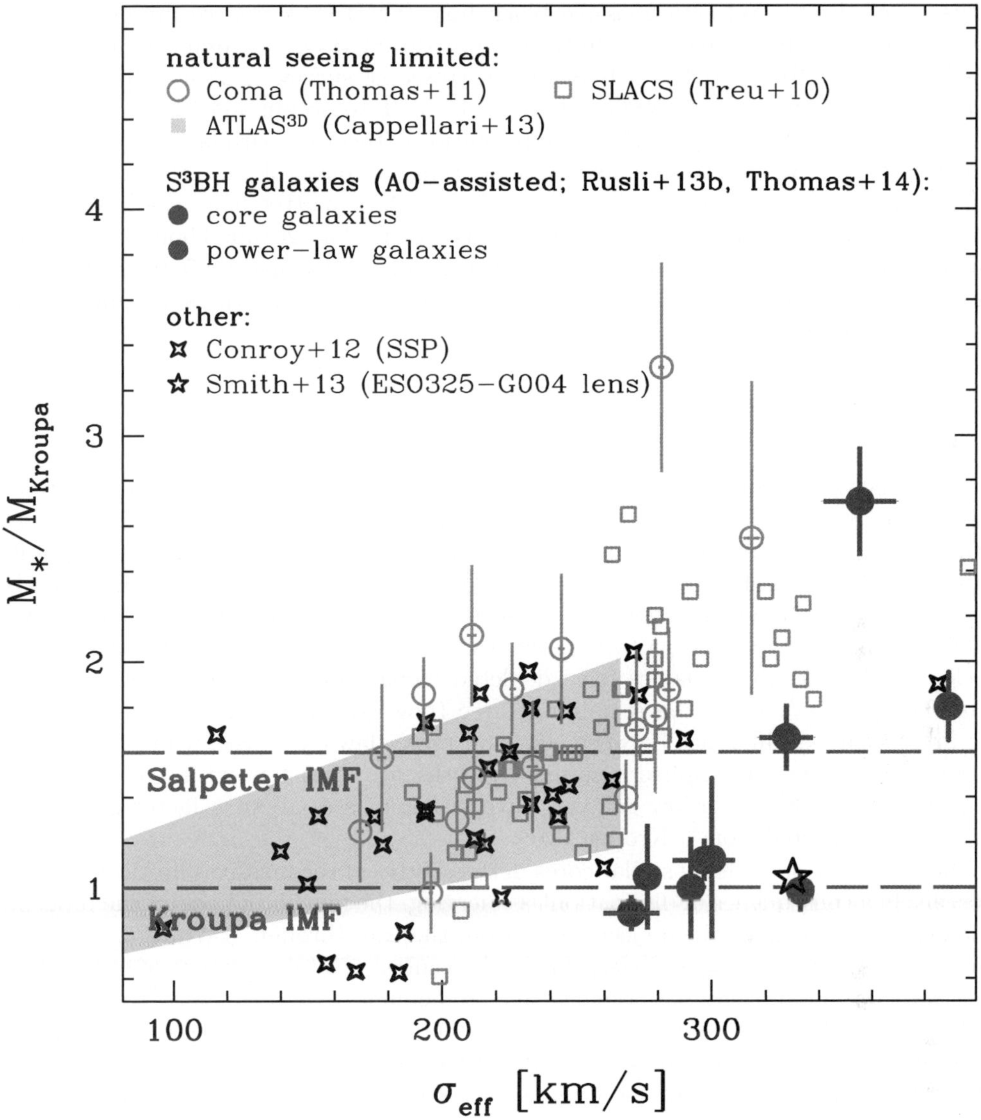

Figure 1. The IMF parameter $M_*/M_{\rm Kroupa}$ as a function of galaxy velocity dispersion $\sigma_{\rm eff}$. Grey: M_* from kinematical and/or lensing data that do not resolve the central SMBH (Treu *et al.* 2010; Thomas *et al.* 2011; Cappellari *et al.* 2013); colored: new M_* from adaptive-optics assisted kinematics in S^{3}BH (Rusli *et al.* 2013b; Thomas *et al.* 2014); open diamonds: M_* from stellar population models with variable IMF constrained from dwarf-star sensitive spectral features (Conroy & van Dokkum 2012); open star: M_* from gravitational lensing (Smith & Lucey 2013).

galaxies seems to be more dwarf dominated than in the Milky-Way (and can be described, e.g., by a Salpeter IMF).

Neither of the two approaches is free of systematics. While direct methods may suffer from not-yet-understood abundance effects on the IMF sensitive spectral features, dynamical and/or lensing masses in general can only determine the total gravitating mass of a galaxy – its decomposition into the contribution of the stars and of the dark matter halo is degenerate and model dependent. So far, indirect IMF measurements from stellar dynamics have been based on kinematical data that do not resolve the sphere of influence of the central supermassive black hole (SMBH). Here, we present new constraints on the IMF in massive ellipticals from the SINFONI Search for Supermassive Black Holes (S^3BH). S^3BH is a K-band spectroscopic survey of about 30 disk and elliptical galaxies, using the SINFONI IFU at the VLT to obtain adaptive-optics assisted central stellar kinematics (≈ 0.15 arcsec resolution). Together with larger-scale kinematical data this allows us to dynamically constrain the mass of the central SMBH, the stellar mass-to-light ratio Υ and the mass in the dark matter halo at the same time, using Schwarzschild orbit superposition models (Rusli *et al.* 2013b; Thomas *et al.* 2014). The high spatial resolution of our data further allows us to constrain Υ in the innermost regions of the galaxies, where any possible degeneracy between the stars and the dark matter halo is least significant.

2. Results

Fig. 1 shows our constraints on the IMF for the subsample of nine high-σ early-type galaxies in S^3BH. This sample is identical with the one described in Thomas *et al.* (2014), except for two galaxies that we here omit for their low σ (< 250 km/s). Fig. 1 shows, as a function of galaxy velocity dispersion σ, the ratio of the dynamically derived stellar mass M_* and a stellar population estimate $M_{\rm Kroupa}$ using the models of Maraston (2005) and assuming a Kroupa IMF. Galaxies from S^3BH are shown by the filled colored circles. Six of the nine galaxies have depleted stellar cores and are shown in red (cf. Rusli *et al.* 2013a), while the remaining three (power-law) galaxies are shown in blue. We find a marked difference between core and power-law ellipticals: while core galaxies, in contrast to the trend implied by previous works, are consistent with a Milky-Way like (e.g. Kroupa) IMF, power-law galaxies have stellar masses about a factor of two times larger than expected from a Kroupa IMF.

The formation of depleted stellar cores in the centres of the massive elliptical galaxies is most likely a consequence of dissipationless merging: the two black holes of the progenitor galaxies form a binary, which ejects stars from the core through gravitational slingshot. This process leaves a unique fingerprint in the central orbital structure of core galaxies: only stars on radial orbits pass the centre close enough to get ejected by the interaction with the SMBH binary, while stars on tangential orbits remain unaffected. Core galaxies are therefore of particular importance, since their unique central orbital structure allows an independent crosscheck of the dynamical models. In Thomas *et al.* (2014) we show that our best-fit dynamical models for the core ellipticals are in very good agreement with the predicted tangential anisotropy in the centres of these galaxies. Higher mass-to-light ratios, as indicative for a bottom-heavy IMF are not only ruled out by the kinematical data, but would also require an orbital structure in contradiction with the core scouring picture.

The bottom-light IMF in our high-σ core ellipticals causes tension with the statistical trend implied by previous studies, from which a Salpeter or even more dwarf dominated IMF had been expected. Previous dynamical studies, however, were based on kinematical data with lower spatial resolution that did not resolve the central SMBH. Our findings suggest that kinematic data with high central spatial resolution may be essential to derive unbiased dynamical stellar mass-to-light ratios. Smith & Lucey (2013) recently also reported evidence for a bottom-light IMF in a nearby, high-σ (lensing) galaxy. The range of the stellar IMF in massive galaxies may therefore be larger than anticipated thus far. The data is also consistent with a non-variable, Kroupa IMF in all galaxies, if part of the central mass in the power-law galaxies is a component of dark matter that is degenerate with the stellar mass distribution at the resolution limit of present-day kinematical observations.

References

Cappellari, M., *et al.*, 2013, *MNRAS*, 432, 1862
Conroy, C. & van Dokkum, P. G., 2012, *ApJ*, 760, 71
Maraston, C., 2005, *MNRAS*, 362, 799
Rusli, S. P., Erwin, P., Saglia, R. P., Thomas, J., Fabricius, M., Bender, R., & Nowak, N., 2013a, *AJ*, 146, 160
Rusli, S. P., *et al.*, 2013b, *AJ*, 146, 45
Smith, R. J. & Lucey, J. R., 2013, *MNRAS*, 434, 1964
Thomas, J., *et al.*, 2011, *MNRAS*, 415, 545
Thomas, J., Saglia, R. P., Bender, R., Erwin, P., & Fabricius, M., 2014, *ApJ*, 782, 39
Treu, T., Auger, M. W., Koopmans, L. V. E., Gavazzi, R., Marshall, P. J., & Bolton, A. S., 2010, *ApJ*, 709, 1195

Galaxy Masses as Constraints of Formation Models
Proceedings IAU Symposium No. 311, 2014
M. Cappellari & S. Courteau, eds.

doi:10.1017/S174392131500335X

X-Ray Measurements of the Mass Profiles in Massive Isolated Elliptical Galaxies

David A. Buote[1], Ewan O'Sullivan[2] and Trevor J. Ponman[3]

[1]Department of Physics and Astronomy, 4129 Frederick Reines Hall, University of California, Irvine, Irvine, CA 92697-4575 USA. email: buote@uci.edu
[2]Harvard-Smithsonian Center for Astrophysics, 60 Garden Street, MS-50 Cambridge, MA 02138, USA. email: eosullivan@cfa.harvard.edu
[3]School of Physics and Astronomy, University of Birmingham, Birmingham, B15 2TT, UK. email: tjp@star.sr.bham.ac.uk

Abstract. We discuss some results for mass profiles that we have obtained from hydrostatic studies of the hot gas in isolated, massive elliptical galaxies.

Keywords. galaxies: elliptical and lenticular, dark matter: galaxies, cD - galaxies: evolution, galaxies: formation, galaxies: fundamental parameters, galaxies: ISM, X-rays: galaxies

1. Introduction

The mass profiles of elliptical galaxies are important probes of galaxy formation and cosmology via measurements of their total baryon fraction and the relationship between halo concentration and virial mass. Only a few elliptical galaxies have detailed mass measurements from X-rays because of the lack of suitable targets in terms of total mass ($\lesssim 10^{13}\ M_{\odot}$), ISM X-ray flux, and relaxed dynamical states appropriate for hydrostatic analysis (e.g., for a review see Buote & Humphrey 2012a). We have identified a sample of 11 (hot) gas-rich isolated elliptical galaxies, most of which were discovered from previously awarded *XMM-Newton* and *Chandra* snapshot programs to search for optimal targets for X-ray studies of mass profiles on the galaxy scale.

2. Hydrostatic Equilibrium

Since the primary tool to obtain mass profiles of elliptical galaxies from X-rays is through the hydrostatic equilibrium approximation, we discuss briefly its validity. Recall that for a given fluid element hydrostatic equilibrium is the balance between the inward force of gravity and the outward force due to thermal gas pressure. One would like to know the importance of correction terms to this equation. Ideally, any non-thermal gas motions could be measured by Doppler shifts and broadening of the many strong emission lines in the X-ray–emitting plasma. This will be possible for the brightest systems with the launch of *Astro-H* in 2015. Kinematic constraints with current data are challenging but suggest that gas motions near the centers of elliptical galaxies are sub-sonic (e.g., Werner *et al.* 2009; Sanders *et al.* 2011).

Although presently one cannot rely on direct kinematic constraints on non-thermal gas motions, there are still reasonable steps that can be taken to mitigate their impact on mass-profile studies (Buote & Humphrey 2012a). First, for the hydrostatic equilibrium approximation to be most accurate, one should focus on those systems that are the most regular in terms of their image morphology. If there is a disturbance, such as from AGN feedback, it is best if the disturbance is confined to a spatial scale much smaller than the region of interest. Next, it is important that the X-ray emission be dominated by that from the hot gas not from background sources. This is especially important in elliptical galaxies (as opposed to galaxy clusters), where sometimes the emission from unresolved discrete sources can masquerade as diffuse gas. The third guideline is to examine whether a reasonable hydrostatic model is able to provide an acceptable description of the X-ray data. While this does not prove hydrostatic equilibrium, it is a necessary consistency check. Finally, until the day arrives when precise constraints on detailed gas kinematics are routinely available, the best means to gauge the reliability of hydrostatic equilibrium is to compare to mass measurements obtained by other methods such as stellar dynamics and gravitational lensing.

Comparisons have been performed between stellar dynamics and X-rays where it was assumed that all differences between the mass profiles obtained by the two techniques were due solely to non-thermal gas motions. A study by Churazov *et al.* (2010) of six galaxies and one galaxy by our team (Humphrey *et al.* 2013) find significant offsets leading to approximately a 30% non-thermal pressure fraction. These results pertain to galaxies at the centers of group-scale halos with total virial masses of a few $\times 10^{13}\ M_{\odot}$. It is unclear whether they hold for smaller, galaxy-scale halos.

3. Isolated Elliptical Galaxy Sample

Here we are interested in the properties of elliptical galaxies outside of any group or cluster. This leads us to focus on isolated elliptical galaxies. Isolation guarantees a galaxy-scale dark matter halo provided the ratio of X-ray to optical luminosity is not too large (otherwise one would have a fossil group). In the most isolated ellipticals the hydrostatic equilibrium approximation is expected to hold most accurately, since the gas has had ample time to settle down from past mergers. Unfortunately, only a few isolated elliptical galaxies with large hot gas halos have been observed previously in X-rays. To remedy this, we constructed an optically selected sample of isolated elliptical galaxies from the LEDA catalog with various criteria. Our total sample consists of 35 galaxies within 100 Mpc. Only 10 were previously observed with deep X-ray observations. We conducted a snapshot campaign with *XMM-Newton* and *Chandra* of the rest of the sample to ascertain their fluxes more accurately. Our survey is now completed, and we have now identified 11 isolated elliptical galaxies with substantial X-ray halos, 8 of which presently have deep *Chandra* observations. We will focus on these systems in this proceedings.

4. Results

4.1. *Spatially Resolved Spectral Analysis*

First we will summarize results we obtained previously for two of the brightest systems in our sample using traditional spatially resolved X-ray spectroscopy (Humphrey *et al.* 2011, 2012). NGC 720 is a well-known object which has been studied extensively in the past. The extended X-ray emission in NGC 1521 was discovered in our snapshot campaign. For these galaxies we also have deep *XMM-Newton* or *Suzaku* observations to complement the *Chandra* data allowing for constraints out close to R_{2500}, which is about 1/4 the virial radius. In Figure 2 of Humphrey *et al.* (2011) we present some representative *Chandra* and *Suzaku* spectra of NGC 720 at different radii. In these spectra one can notice that at the largest radius the background exceeds the source clearly indicating that accurate background subtraction is essential for constraining the gas properties. Nevertheless, even in the outermost aperture the gas properties can be constrained because the spectral "bump" near 1 keV is visible. This feature arises from the unresolved emission from the strong, temperature-sensitive Fe-L shell lines. Indeed, the well-constrained temperature and density profiles of the hot gas of NGC 720 obtained by fitting these *Chandra* and *Suzaku* spectra are shown in Figure 3 of Humphrey *et al.* (2011). To constrain the mass profile, we fitted these temperature and density profiles with hydrostatic models. Shown in that Figure 3 are the results of fitting a model without any dark matter halo. The fit is obviously very poor and is formally excluded at very high significance; i.e., about 20σ using the Bayesian evidences.

Hydrostatic equilibrium can be applied in many different ways. We find an entropy-based procedure is particularly effective since it allows us to easily enforce the additional constraint of convective stability (e.g., Buote & Humphrey 2012a). We begin by assuming parameterized forms for the entropy and mass profiles which are input into the equation of hydrostatic equilibrium. Then we compute the gas density and temperature profiles which are fitted to the observations. We also mention that provided the system is approximately a triaxial ellipsoid, we have shown previously that the assumption of spherical symmetry translates into only small errors in the derived properties (Buote & Humphrey 2012b). Finally, we note that we use the NFW profile for the dark matter halo in our default analysis while a 2-break power-law model (with a constant) is used for the entropy profile. For NGC 720, e.g., we have 9 free parameters constrained by 26 data points.

For both NGC 720 and NGC 1521 we obtain very tight constraints on the mass within the region where we have data (see Figure 6 of Humphrey *et al.* 2011 and Figure 7 of Humphrey *et al.* 2012). When we evaluate our models self-consistently out to the virial radius the error bars increase but we clearly find a total masses below $10^{13}\ M_{\odot}$ indicative of a galaxy-scale halo in each case. We also find good constraints on the global baryon fractions that are very consistent with the cosmic value. These results provide interesting evidence that the so-called "Missing Baryons" may exist in the outer parts of galaxy halos as hot gas – at least for massive ellipticals. It also shows that any non-gravitational heating did not push much gas outside of the virial radius.

Finally, we mention that we have constructed detailed error budgets for our measured parameters considering how different analysis choices affect our results for both galaxies. For NGC 720 the largest effect at the virial radius is the choice of dark matter profile. If instead of NFW we use a cored logarithmic potential (equivalent to a β model when the gas is isothermal) then at the virial radius the baryon fraction is shifted down by 0.11 to a value about 0.05. This alternative model, where the dark matter density $\sim r^{-2}$ at the virial radius, can be tested against the NFW model. Examining the bayesian evidences we find that NFW is preferred over the cored logarithmic potential by about 4σ.

4.2. *Narrow-Band Surface Photometry*

It is not always possible or even necessary to conduct detailed spectral fitting in each spatial bin to extract the global properties of the hot gas. In Humphrey & Buote (2013) we have studied the utility of narrow-band X-ray photometry for studies of the hot gas in elliptical galaxies and galaxy clusters. In essence, one is exploiting a tradeoff in spectral degrees of freedom for spatial degrees of freedom – ideal for high resolution *Chandra* data. This approach also has the significant advantage of typically having a far simpler background subtraction involving essentially constant components. We have applied this method to the *Chandra* data of 8 of our isolated elliptical galaxies using the following 3 energy bands: 0.5-0.9, 0.9-2.0, and 2.0-7.0 keV. In Figure 1 we show a fit to one of the galaxies in our sample. Notice the hard band, shown in green, reaches the constant

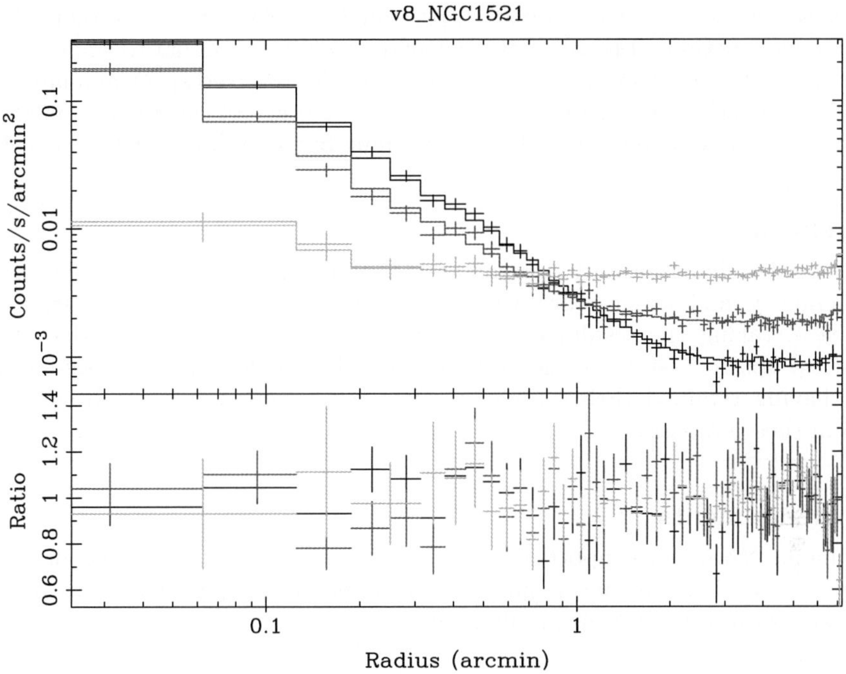

Figure 1. Galaxy + background fitted jointly to the 3-band Chandra surface brightness profile of NGC 1521: soft band (black, 0.5-0.9 keV), medium band (red, 0.9-2.0 keV) and hard band (green, 2.0-7.0 keV). The bottom panel shows the ratio data/model.

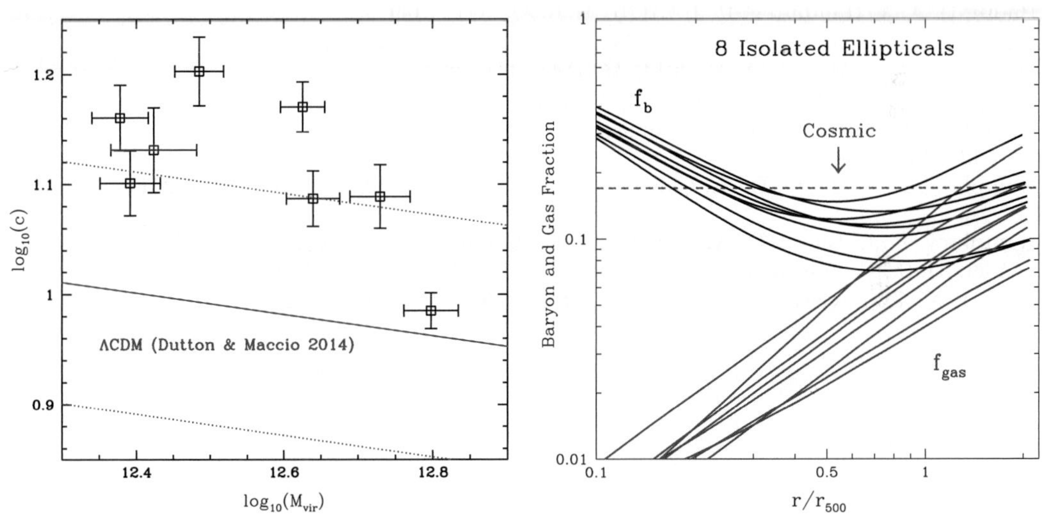

Figure 2. (*Left Panel*) Concentration-mass relation where each quantity is defined at the virial radius ($\Delta \sim 101$). (*Right Panel*) Radial profiles of the gas (blue) and baryon (black) fractions.

background level very quickly. But in softer bands the gas dominates the emission in the stellar region and does not reach the background level until a radius of about 50 kpc.

We display the key results for the mass profiles in Figure 2. In the left plot we show the concentrations and virial masses. A simple average of the virial masses of the 8 galaxies gives a value of $4 \times 10^{12}\ M_{\odot}$. The measurements lie systematically above the ΛCDM concentration-mass relation from Dutton & Macciò (2014) which may be expected since our galaxies were selected to be isolated, presumably early forming systems. As for the baryon and gas fractions (Figure 2), the behavior is also very similar to what we found for NGC 720 and NGC 1521 from the traditional spatially resolved spectral analysis. At the virial radius – which is about twice r_{500} – a simple average of the 8 values yields a value for the baryon fraction consistent the cosmic value but with large scatter ($\sigma = 0.07$).

5. Conclusions

We emphasize that the results presented for the narrow-band surface photometry are still preliminary, and we are currently constructing detailed systematic error budgets. Nevertheless, taking the results presented in §4.2 at face value we find that there is no core-cusp controversy for these systems, and these galaxies are consistent with, on average, having retained their cosmic fraction of baryons; i.e., no "Missing Baryons" problem in these systems.

Acknowledgements

D.A.B. thanks NASA for support. We thank Philip Humphrey for early contributions to this project, especially for developing an initial version of the software to perform the narrow-band surface-brightness-profile fitting.

References

Buote, D. A. & Humphrey, P. J. 2012a, in *Astrophysics and Space Science Library*, Vol. 378, Astrophysics and Space Science Library, ed. D.-W. Kim & S. Pellegrini, 235
Buote, D. A. & Humphrey, P. J. 2012b, *MNRAS*, 421, 1399
Churazov, E., Tremaine, S., Forman, W., *et al.* 2010, *MNRAS*, 404, 1165
Dutton, A. A. & Macciò, A. V. 2014, *MNRAS*, 441, 3359
Humphrey, P. J. & Buote, D. A. 2013, *MNRAS*, 436, 2879
Humphrey, P. J., Buote, D. A., Brighenti, F., Gebhardt, K., & Mathews, W. G. 2013, *MNRAS*, 430, 1516
Humphrey, P. J., Buote, D. A., Canizares, C. R., Fabian, A. C., & Miller, J. M. 2011, *ApJ*, 729, 53
Humphrey, P. J., Buote, D. A., O'Sullivan, E., & Ponman, T. J. 2012, *ApJ*, 755, 166
Sanders, J. S., Fabian, A. C., & Smith, R. K. 2011, *MNRAS*, 410, 1797
Werner, N., Zhuravleva, I., Churazov, E., *et al.* 2009, *MNRAS*, 398, 23

Galaxy Masses as Constraints of Formation Models
Proceedings IAU Symposium No. 311, 2014
M. Cappellari & S. Courteau, eds.

doi:10.1017/S1743921315003361

Reconstructing the mass assembly history with kinematics and nuclear light profiles

Davor Krajnović

Leibniz-Institut für Astrophysik Potsdam (AIP), An der Sternwarte 16, D-14482 Potsdam, Germany. email: dkrajnovic@aip.de

Abstract. In this contribution I show that by combining imaging and integral-field spectroscopy it is possible to unravel the internal structure of galaxies. In particular, I will present the photometric and kinematic evidence for discs linking them with stellar angular momentum content of early-type galaxies. Furthermore, I show that the existence of both fast rotators with core and slow rotators with core-less nuclear light profiles challenges the standard formation scenarios for fast and slow rotators and suggests new pathways of mass assembly.

Keywords. galaxies: elliptical and lenticular, cD - galaxies: evolution - galaxies: formation

1. Introduction

In the seminal paper (Davies *et al.* 1983), Roger Davies and collaborators showed that early-type galaxies (ETGs) have divers kinematics, ranging from fainter and rotational supported to brighter and dispersion dominated systems. This picture is effectively still with us (e.g. Cappellari *et al.* 2007; Emsellem *et al.* 2011), although the focus has somewhat changed. In particular, the usage of integral field spectrographs and the analysis of kinematic maps enable a more detailed analysis of the internal structure of galaxies.

The ATLAS^{3D} Project (Cappellari *et al.* 2011) is a survey of morphologically selected ETGs observable with the SAURON integral-field spectrograph (Bacon *et al.* 2001) mounted on the William Herschel Telescope. The galaxies selected were brighter than -21.5 magnitude in the 2MASS K_s band and within 42 Mpc in order to properly exploit SAURON's short wavelength range. Out of about 900 such galaxies, 260 ETGs were selected, based purely on their morphology. The sample galaxies (or their subsets) were observed with SAURON, IRAM 30m single-dish telescope, Carma Array, Westerbork Radio Synthesis Array, the Isaac Newton Telescope (INT) and the MegaCam at the Canada-France-Hawaii Telescope. In this contribution, I will review the analysis of the SAURON kinematics maps and of ground and space based imaging, looking for evidence that can be used to constrain the assembly processes of ATLAS^{3D} galaxies.

2. Regularity of kinematics and the shape of early-type galaxies

ETGs can be divided into fast and slow rotators (Emsellem *et al.* 2007), a division which to the first order follows the brightness (or mass) (Emsellem *et al.* 2011). The distinction between fast and slow rotators is empirically defined in the space of the specific angular momentum calculated within one effective radius, λ_R, and the observed ellipticity (also estimated within one effective radius), such that $\lambda_R = 0.31 \times \sqrt{\epsilon}$. This empirical division follows another empirical finding that the velocity maps of ETGs either exhibit regular or non-regular velocity features (Krajnović *et al.* 2011). The regularity of a velocity map is measured using `kinemetry` (Krajnović *et al.* 2006), a generalisation of the method of isophotometry (e.g. ?) to higher order moments of the line-of-sight velocity distribution (LOSVD). For the velocity maps, the best fitting ellipse is parameterised by

$V = V_{rot} \cos\theta$, where V_{rot} is the amplitude of rotation and θ is the eccentric anomaly. Deviations from the cosine law, parameterised with the higher order Fourier harmonics, can be quantitatively measured (Krajnović *et al.* 2008) and used to separate regular velocity maps, such as the 'spider' diagrams of the discs galaxies, from non-regular, complex looking velocity maps, often containing kinematically distinct components, strong velocity twists or even no mean rotation. Krajnović *et al.* (2011) showed that more than 80% of ATLAS^{3D} galaxies have regular velocity maps, which, using the criteria for distinction between fast and slow rotators, translates into 86% of fast rotators.

The velocity maps can provide one further evidence of the remarkable regular structure found in the majority of ETGs. Following Franx *et al.* (1991), we can define the *kinematic misalignment angle*, Ψ, as the angle between the photometric major axis and the orientation of the velocity map. The kinematic misalignment angle is a powerful tool to constrain the shape of galaxies, as when Ψ is different from zero, the galaxy can not have an axisymmetric shape. In Krajnović *et al.* (2011), we measured the photometric position angle at large radii (~ 2.5 effective radii) and the kinematic position angle within the SAURON velocity maps (typically within one effective radius) and found that galaxies with regular velocity maps (fast rotators) are typically nearly axisymmetric systems (their $\Psi \lesssim 5^{o}$). On the other hand, galaxies with non-regular velocity maps (slow rotators) show strong kinematic misalignments (up to 90°), suggesting many of them are triaxial systems.

3. Photometric and kinematics evidence for discs

Are the regular velocity maps and the axisymmetric structure of the vast majority of ETGs a consequence of hidden discs? What fraction of the ETG population actually can be considered disc dominated galaxies? The first step in answering these questions can be achieved by analysis of images. In Krajnović *et al.* (2013a), we decomposed the surface brightness profiles, extracted from the SDSS (Abazajian *et al.* 2009) and INT (Scott *et al.* 2013) images, by fitting two component models consisting of a **?** fitting function representing the bulge, and an exponential function, representing a disc. From this decomposition we excluded galaxies with (obvious) bars, as they would require a three component model and bars live in discs. The results show a general correspondence between fast/slow division and the existence of discs: discs are typically found and, actually, dominate fast rotators. Light profiles of some slow rotators, however, suggest embedded and small discs could be found in this population as well. A remarkable result is that the distribution of the disc-to-total light ratios (D/T) in fast rotators covers essentially the full range, from small disc contributions to fast rotators which are essentially lacking a bulge-like component†.

Due to the deprojection effects, finding discs in the photometry is essentially degenerate (Rix & White 1990; Gerhard & Binney 1996). It should be, however, more likely to determine the internal structure from kinematics. To demonstrate this, I will here make use of the third Gauss-Hermite coefficient, h_3, in the parametric expansion of the LOSVD (e.g. van der Marel & Franx 1993). This parameters measure the asymmetric deviations of the LOSVD from a Gaussian and it was shown to tightly anti-correlate with the mean velocity of the same LOSVD (e.g. Bender *et al.* 1994). This anti-correlation is a direct consequence of the existence of fast moving stars, as one would expect in an embedded disc. In galaxies which have a significant disc component or have a single Sérsic profile of a low (< 3) index, the anti-correlation produces a specific shape in the $h_3 - V/\sigma$, as shown on Fig. 1 (top left). The same general shape is found in fast rotators, but a very different

† A number of light profiles of fast rotators are best fitted with only a single Sérsic function, but the Sérsic index n is typical small ($< 2 - 3$).

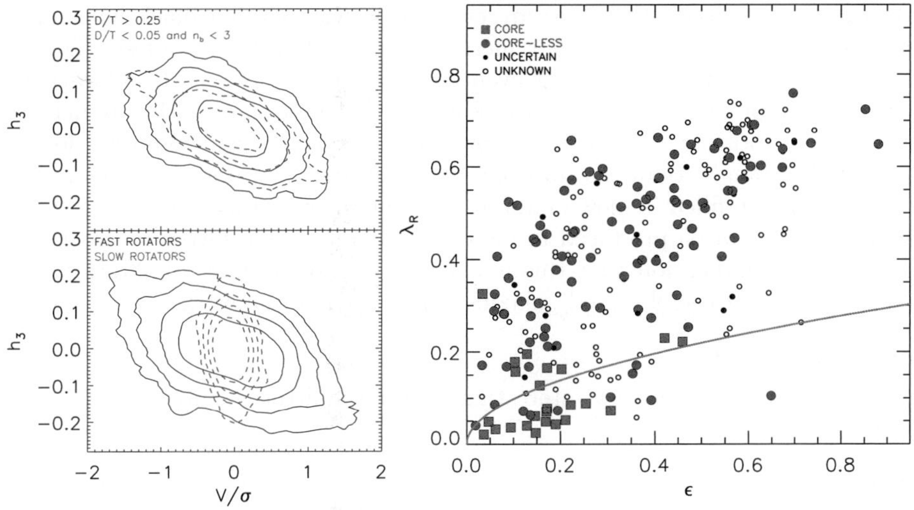

Figure 1. *Left:* Local $h_3 - V/\sigma$ relation for all ATLAS^{3D} galaxy spectra with $\sigma > 120$ km/s and an error on $h_3 < 0.05$. Top panel shows the contours for galaxies with significant disc contribution or galaxies with single Sérsic profiles of a low Sérsic index (dashed). Bottom panel shows the distribution for fast and slow rotators (dashed). *Right:* Angular momentum vs ellipticity for the ATLAS^{3D} sample. The green solid line separates fast (above) and slow (below) rotators. Various symbols show the properties of nuclear light profiles, as stated on the legend.

shape (no anti-correlation) is seen for slow rotators (Fig. 1; bottom left). The implication is simple: fast rotators are indeed disc dominated systems. Discs, whether large or small and embedded, are present in the majority of ETGs, and support a new classification scheme (Cappellari *et al.* 2011), which should substitute the classical Hubble diagram.

4. Angular momentum and nuclear light profies

Nuclear surface brightness profiles of massive galaxies are known to show a bend downwards or a deficit of light with respect to a Sérsic fit at large radii (e.g. Lauer *et al.* 1995; Graham *et al.* 2003). Galaxies with such profiles are often said to have *cores*, while galaxies whose profiles stay consistent with the outer Sérsic fit or even show a nuclear excess of light, we will call here *core-less*. Core galaxies are typically more massive (Faber *et al.* 1997), as much as are slow rotators and a natural question is whether there is a clear correspondence between slow rotators and cores. In Krajnović *et al.* (2013b), we investigated this issue by matching the observations in the HST archives with ATLAS^{3D} galaxies. To a first approximation, fast rotators are core-less galaxies, while slow rotators are cores, but there are some exceptions (Fig 1, right). Notably there is a population of fast rotators with cores. They are kinematically consistent with being fast rotators with large discs (e.g. regular kinematics, $\Psi \approx 0^{\circ}$, signifiant $h_3 - V/\sigma$ anti-correlation), but have $\lambda_R < 0.25$ and they are not far from the line separating fast and slow rotators. There are also some core-less slow rotators, while for a number of slow rotators their nuclear profiles are not know. Even if one removes some special cases among slow rotators†, it is obvious there is no 1:1 correspondence between cores and slow rotators and core-less and fast rotators.

5. Discussion

What are core-less slow rotators and how are they made? How is it possible to make, or preserve, a core in a disk dominated fast rotator? These types of galaxies clearly require

† For example, recent mergers, or galaxies made up of two counter-rotating discs.

special formation paths. If the cores are created by the merging black holes of similar masses (brought together by a major merger event), while the core-less (light excess) profiles are made in nuclear starburst events, a fine tuning between the duration of the starburst and the binary evolution is necessary for the creation of both core fast rotators or core-less slow rotators. Creating the cores in fast rotators will require a longer evolution of the BH binary than the duration of the starburst, while the cores in the slow rotators need to be refilled by new material after the black holes are merged. This means that slow rotators can be produced in dissipative, while fast rotators in dissipation-less merges. There are indication that this is indeed the case (Bois *et al.* 2011; Naab *et al.* 2014), but simulating the observed galaxies still requires some work.

Acknowledgements

I thank the organisers for a very interesting meeting and the members of ATLAS^{3D} for making this project a success.

References

Abazajian, K. N., Adelman-McCarthy, J. K., Agüeros, M. A., *et al.* 2009, *ApJS*, 182, 543
Bacon, R., Copin, Y., Monnet, G., Miller, B. W., Allington-Smith, J. R., Bureau, M., Carollo, C. M., Davies, R. L., Emsellem, E., Kuntschner, H., Peletier, R. F., Verolme, E. K., & de Zeeuw, P. T., 2001, *MNRAS*, 326, 23
Bender, R., Saglia, R. P., & Gerhard, O. E., 1994, *MNRAS*, 269, 785
Bois, M., Emsellem, E., Bournaud, F., Alatalo, K., Blitz, L., *et al.* 2011, *MNRAS*, 416, 1654
Cappellari, M., Emsellem, E., Bacon, R., Bureau, M., Davies, R. L., de Zeeuw, P. T., *et al.* 2007, *MNRAS*, 379, 418
Cappellari, M., Emsellem, E., Krajnović, D., McDermid, R. M., Scott, N., Verdoes Kleijn, G. A., Young, L. M., Alatalo, K., *et al.* 2011, *MNRAS*, 413, 813
Cappellari, M., Emsellem, E., Krajnović, D., McDermid, R. M., Serra, P., Alatalo, K., Blitz, L., Bois, M., *et al.* 2011, *MNRAS*, 416, 1680
Davies, R. L., Efstathiou, G., Fall, S. M., Illingworth, G., & Schechter, P. L., 1983, *ApJ*, 266, 41
Emsellem, E., Cappellari, M., Krajnović, D., Alatalo, K., Blitz, L., *et al.* 2011, *MNRAS*, 414, 888
Emsellem, E., Cappellari, M., Krajnović, D., van de Ven, G., *et al.* 2007, *MNRAS*, 379, 401
Faber, S. M., Tremaine, S., Ajhar, E. A., Byun, Y.-I., Dressler, A., *et al.* 1997, *AJ*, 114, 1771
Franx, M., Illingworth, G., & de Zeeuw, P. T., 1991, *ApJ*, 383, 112
Gerhard, O. E. & Binney, J. J., 1996, *MNRAS*, 279, 993
Graham, A. W., Erwin, P., Trujillo, I. & Asensio Ramos, A. 2003, *AJ*, 125, 2951
Jedrzejewski, R. I., 1987, *MNRAS*, 226, 747
Krajnović, D., Cappellari, M., de Zeeuw, P. T., & Copin, Y., 2006, *MNRAS*, 366, 787
Krajnović, D., Bacon, R., Cappellari, M., Davies, R. L., *et al.* 2008, *MNRAS*, 390, 93
Krajnović, D., Emsellem, E., Cappellari, M., Alatalo, K., Blitz, L., *et al.* 2011, *MNRAS*, 414, 2923
Krajnović, D., Alatalo, K., Blitz, L., Bois, M., Bournaud, F., *et al.* 2013a, *MNRAS*, 432, 1768
Krajnović, D., Karick, A. M., Davies, R. L., Naab, T., Sarzi, M., *et al.* 2013b, *MNRAS*, 433, 2812
Lauer, T. R., Ajhar, E. A., Byun, Y.-I., Dressler, A., Faber, S. M., *et al.* 1995, *AJ*, 110, 2622
Naab, T., Oser, L., Emsellem, E., Cappellari, M., Krajnović, D., *et al.* 2014, *MNRAS*, 444, 3357
Rix, H. & White, S. D. M., 1990, *ApJ*, 362, 52
Scott, N., Cappellari, M., Davies, R. L., Kleijn, G. V., Bois, M., *et al.* 2013, *MNRAS*, 432, 1894
Sérsic, J. L., 1968, Atlas de galaxias australes. Cordoba, Argentina: Observatorio Astronomico, 1968
van der Marel, R. P. & Franx, M., 1993, *ApJ*, 407, 525

Galaxy Masses as Constraints of Formation Models
Proceedings IAU Symposium No. 311, 2014
M. Cappellari & S. Courteau, eds.

doi:10.1017/S1743921315003373

Connection between dynamically derived IMF normalisation and stellar populations

Richard M. McDermid[1,2]

[1]Department of Physics and Astronomy, Macquarie University, Sydney NSW 2109, Australia
[2]Australian Astronomical Observatory, PO Box 915, Sydney NSW 1670, Australia
email: richard.mcdermid@mq.edu.au

Abstract. In this contributed talk I present recent results on the connection between stellar population properties and the normalisation of the stellar initial mass function (IMF) measured using stellar dynamics, based on a large sample of 260 early-type galaxies observed as part of the ATLAS^{3D} project. This measure of the IMF normalisation is found to vary non-uniformly with age- and metallicity-sensitive absorption line strengths. Applying single stellar population models, there are weak but measurable trends of the IMF with age and abundance ratio. Accounting for the dependence of stellar population parameters on velocity dispersion effectively removes these trends, but subsequently introduces a trend with metallicity, such that 'heavy' IMFs favour lower metallicities. The correlations are weaker than those found from previous studies directly detecting low-mass stars, suggesting some degree of tension between the different approaches of measuring the IMF. Resolving these discrepancies will be the focus of future work.

Keywords. galaxies: elliptical and lenticular, cD - galaxies: evolution - galaxies: stellar content

1. Introduction

A universal form for the stellar initial mass function (IMF) has been an actively debated topic for many decades, having far-reaching consequences for our understanding of the star formation process, and more generally, of how to interpret the integrated light we observe coming from unresolved stellar populations in galaxies and star clusters. On one hand, the environment (i.e. physical conditions) for star formation and the evolutionary processes at work can vary significantly from galaxy to galaxy, suggesting that a non-universal IMF should not be ruled out. On the other, until recently there was little compelling evidence that could sufficiently isolate the properties of the IMF in such unresolved systems, and thus constrain its approximate functional form.

A recent breakthrough was the convincing detection by van Dokkum & Conroy (2010) of an over-abundance of low-mass stars compared to fiducial IMFs. Subsequently, strong evidence for a non-universal IMF has emerged from the application of various independent techniques, including gravitational lensing (Auger *et al.* 2010), stellar dynamical modelling (Cappellari *et al.* 2012, 2013b; Tortora *et al.* 2013; Conroy *et al.* 2013), and spectral synthesis (van Dokkum & Conroy 2010, 2011; Spiniello *et al.* 2012; La Barbera *et al.* 2013; Ferreras *et al.* 2013; Spiniello *et al.* 2014).

Here we use the IMF derived via the stellar dynamical modelling of Cappellari *et al.* (2012) to explore whether the IMF - a key property of any stellar population - shows any measurable systematic variations with galaxy stellar population parameters, as derived via line-strengths and single stellar population models. This summarises work presented in McDermid *et al.* (2014) and McDermid *et al.* (2015).

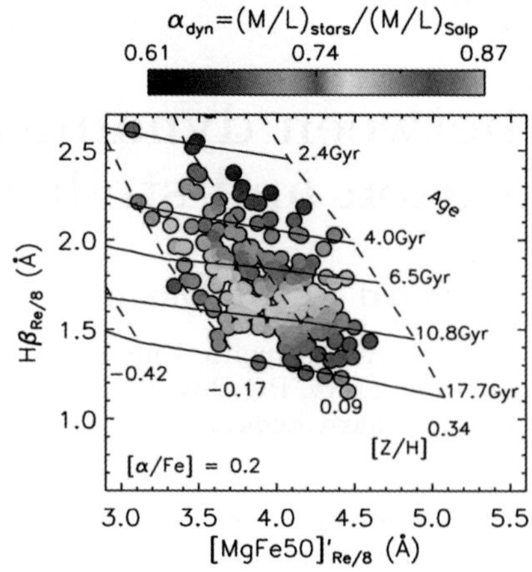

Figure 1. Lick index Hβ versus the combined index [MgFe50]$'$, both measured within an aperture of one eighth of an effective radius. The colour scale indicates the values of the IMF parameter, $\alpha_{\rm dyn}$, which have been adaptively smoothed using a locally-weighted regression technique to show underlying trends in the distribution. A grid of SSP model predictions from Schiavon (2007) is shown for a super-solar abundance ratio of $[\alpha/\text{Fe}]= 0.2$, as indicated in the lower left of the plot. Solid lines indicate lines of constant age; dashed lines show constant metallicity.

2. Methods

We quantify the mass normalisation of the IMF using the IMF parameter $\alpha_{\rm dyn} \equiv (M/L)_{\rm stars}/(M/L)_{\rm Salp}$, such that $\alpha_{\rm dyn} > 1$ implies an IMF 'heavier' than the fiducial unimodal power law of the form: $\zeta(m) \propto m^{-2.35}$ (Salpeter 1955). The nominator of this ratio is derived from fitting dynamical models to the observed integral-field stellar kinematics. These models account for differing orbital anisotropy between galaxies, and implicitly reproduce the shape and profile of the observed surface brightness to high accuracy. Additionally, the presence of dark matter was explored using various general prescriptions for the halo mass profile, including contracted haloes. In this way, the inferred mass-to-light ratio entering into the $\alpha_{\rm dyn}$ parameter is dynamical measure of the stars *only*. The dynamical models are described in full in Cappellari *et al.* (2013a).

The denominator of the $\alpha_{\rm dyn}$ parameter is derived from fitting stellar population models to the ATLAS$^{\rm 3D}$ optical spectroscopy, as described in McDermid *et al.* (2015). Using the penalised pixel fitting method of Cappellari & Emsellem (2004), the spectra are fitted with a (positive) regularised linear combination of the stellar population models of Vazdekis *et al.* (2012), fitting for age and metallicity as free parameters, and holding the IMF of the models fixed at Salpeter. With this approach, all the information in the spectrum is used to constrain the spread in possible ages and metallicities, and thus accounts for variations in star formation history between different galaxies.

Finally, the stellar population parameters are measured using line strengths and the single stellar population (SSP) models of Schiavon (2007). These models have variable abundance ratios, encapsulated in the ratio of α-elements to iron, $[\alpha/\text{Fe}]$, in addition to age and total metallicity, [Z/H]. The measurements are taken from McDermid *et al.* (2015), from which we use the central apertures covering $R_e/8$ to be consistent with existing spectral IMF studies.

3. Results

Fig. 1 presents the averaged distribution of $\alpha_{\rm dyn}$ values on the plane of Hβ versus the combined index [MgFe50]$'$ of Kuntschner *et al.* (2010), over plotted with the Schiavon

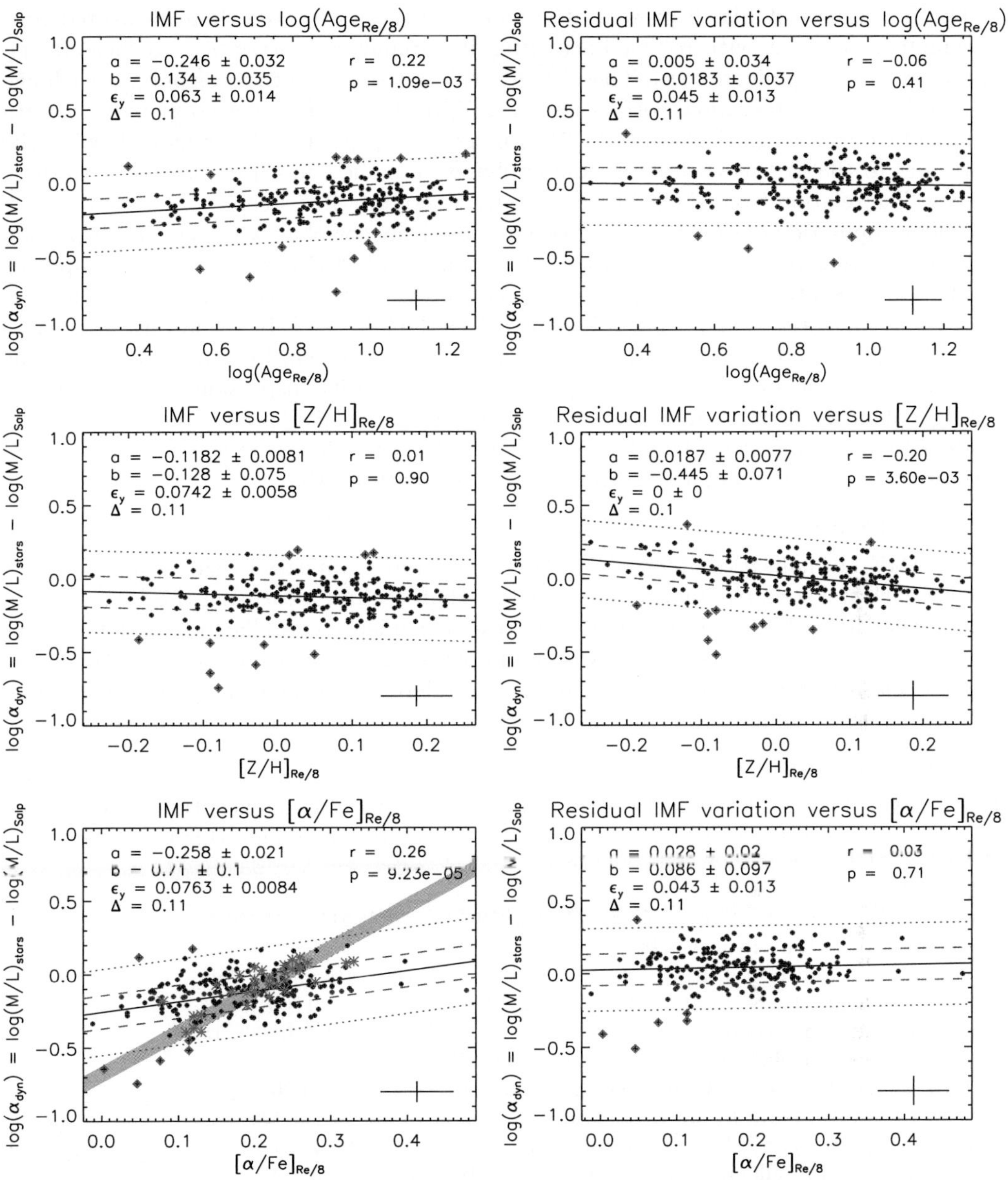

Figure 2. *Left:* Relations of IMF parameter (y-axis) with (top to bottom) age, total metallicity and [α/Fe], showing in each panel the linear fit parameters (upper left) and correlation coefficient (r) and significance (p, upper right). The bottom panel also shows points from Conroy & van Dokkum (2012) in magenta, over-plotted with a band indicating the linear fit obtained using the same algorithm as for the ATLAS^{3D} points. *Right:* Equivalent plots to left column, but after the co-dependence on velocity dispersion has been removed. This removes the trends with age and [α/Fe], but a trend with metallicity becomes apparent.

model grid for [α/Fe]= 0.2. There is a trend of larger values of the IMF parameter (corresponding to 'heavier' IMFs) towards weaker Hβ (older), and stronger [MgFe50]$'$ (metal-rich) objects. The trend, however, is not monotonic, with lower metallicity objects also showing relatively high IMF parameter values, on average.

The left panels of Fig. 2 presents the trends of $\alpha_{\rm dyn}$ with the three measured SSP parameters, together with robust linear fits, fit parameters, correlation coefficients (r), and correlation significance (p). These show weak but significant trends of $\alpha_{\rm dyn}$ with age and [α/Fe], and no measured correlation with [Z/H]. For reference, we over-plot the data points of Conroy & van Dokkum (2012) in magenta, together with a region showing the slope and 1σ (68%) scatter derived from applying the same line-fitting procedure to these points. The relation from the ATLAS$^{\rm 3D}$ points is notably less steep.

Stellar population parameters are known to positively correlate with velocity dispersion (Thomas *et al.* 2005; Graves *et al.* 2009; McDermid *et al.* 2015), as does $\alpha_{\rm dyn}$ (Cappellari *et al.* 2013b). Smith (2014) has shown that accounting for this additional dependency can change the interpretation of what primarily drives the inferred IMF variations. To explore this, the right-hand panels of Fig. 2 show the residual correlations after removing the dependency of the stellar population parameters on velocity dispersion. This is done by fitting a plane to the three-dimensional distribution of $\alpha_{\rm dyn}$, SSP parameters, *and* velocity dispersion simultaneously. This indeed removes the dependency on age and [α/Fe], and interestingly implies a *negative* correlation of $\alpha_{\rm dyn}$ with [Z/H].

4. Conclusions

The IMF normalisation inferred via stellar dynamics appears to show weaker correlations with the stellar population parameters than studies measuring the IMF directly from stellar light. Moreover, accounting for the dependence of stellar population parameters with velocity dispersion removes the apparent trends with age and [α/Fe], while introducing a trend with [Z/H]. Resolving the differences of IMF inferences from independent methods remains a challenge for future work in this field.

Acknowledgements

It is a great pleasure to thank the organisers and attendees for an exceptionally enjoyable conference, and to my colleagues in the ATLAS$^{\rm 3D}$ team for making this work possible.

References

Auger, M. W., Treu, T., Bolton, A. S., *et al.* 2010, *ApJ*, 724, 511
Cappellari, M. & Emsellem, E. 2004, *PASP*, 116, 138
Cappellari, M., McDermid, R. M., Alatalo, K., *et al.* 2012, *Nature*, 484, 485
Cappellari, M., Scott, N., Alatalo, K., *et al.* 2013, *MNRAS*, 432, 1709
Cappellari, M., McDermid, R. M., Alatalo, K., *et al.* 2013, *MNRAS*, 432, 1862
Conroy, C. & van Dokkum, P. G. 2012, *ApJ*, 760, 71
Conroy, C., Dutton, A. A., Graves, G. J., *et al.* 2013, *ApJL*, 776, L26
Ferreras, I., La Barbera, F., de la Rosa, I. G., *et al.* 2013, *MNRAS*, 429, L15
Graves, G. J., Faber, S. M., & Schiavon, R. P. 2009b, *ApJ*, 698, 1590
Kuntschner, H., Emsellem, E., Bacon, R., *et al.* 2010, *MNRAS*, 408, 97
La Barbera, F., Ferreras, I., Vazdekis, A., *et al.* 2013, *MNRAS*, 433, 3017
McDermid, R. M., Cappellari, M., Alatalo, K., *et al.* 2014, *ApJL*, 792, LL37
McDermid, R. M., Alatalo, K., Blitz, L., *et al.* 2015, *MNRAS*, in press. arXiv:1501.03723
Salpeter, E. E. 1955, *ApJ*, 121, 161
Schiavon, R. P. 2007, *ApJS*, 171, 146
Smith, R. J. 2014, *MNRAS*, 443, L69
Spiniello, C., Trager, S. C., Koopmans, L. V. E., & Chen, Y. P. 2012, *ApJL*, 753, L32
Spiniello, C., Trager, S., Koopmans, L. V. E., & Conroy, C. 2014, *MNRAS*, 438, 1483
Thomas, D., Maraston, C., Bender, R., & Mendes de Oliveira, C. 2005, *ApJ*, 621, 673
Tortora, C., Romanowsky, A. J., & Napolitano, N. R. 2013, *ApJ*, 765, 8
van Dokkum, P. G. & Conroy, C. 2010, *Nature*, 468, 940
van Dokkum, P. G. & Conroy, C. 2011, *ApJL*, 735, L13
Vazdekis, A., Ricciardelli, E., Cenarro, A. J., *et al.* 2012, *MNRAS*, 424, 157

Galaxy Masses as Constraints of Formation Models
Proceedings IAU Symposium No. 311, 2014
M. Cappellari & S. Courteau, eds.

doi:10.1017/S1743921315003385

ATLAS^{3D} Stellar Population Gradients

Harald Kuntschner

European Southern Observatory, Karl-Schwarzschild-Strasse 2, 85748 Garching, Germany.
email: hkuntsch@eso.org

Abstract. We present stellar population gradients of early-type galaxies from the ATLAS^{3D} survey: a complete, volume-limited multi-wavelength survey of 260 early-type galaxies in the local 42 Mpc volume. Using emission-corrected spectra integrated within elliptical annuli we measure line-strength indices and apply single stellar population (SSP) models to derive SSP-equivalent values of stellar age, metallicity, and alpha enhancement as function of radius. For all galaxies we derive basic linear stellar population gradients versus radius $\log(R/R_e)$. These gradients are examined on their own and versus three mass-sensitive parameters: K-band luminosity M_K, velocity dispersion within one effective radius $\log \sigma_e$, and our dynamical mass M_{JAM}. We find a correlation between positive age gradients (younger centre) and steeper negative metallicity gradients with a Spearman rank correlation coefficient of -0.46 and a significance of 7.65×10^{-15}. Furthermore, we find a robustly estimated mean metallicity gradient of $\Delta[Z/H] = -0.37 \pm 0.01$ for the sample with a significant trend for more massive galaxies to have shallower profiles. While there is no *clear* distinction between fast and slow rotators or signs of environmental influence, we do detect a significantly larger range of [Z/H]-gradients towards low mass galaxies.

Keywords. galaxies: elliptical and lenticular, cD - galaxies: evolution - galaxies: formation

1. Introduction

The spatial distribution of stellar population properties (e.g. age, metallicity and more detailed chemical composition information) in galaxies holds important clues to the formation and evolution mechanisms which shaped these galaxies. Competing formation scenarios and the physical processes involved predict significantly different radial distributions. Thereby, offering the opportunity to use observed stellar population gradients in galaxies as a discriminating tool for the various formation and evolution scenarios.

The aim of this contribution is to provide new insights into stellar population gradients by making use of the high S/N IFU observations of a complete, magnitude limited sample of 260 nearby ETGs in the ATLAS^{3D} survey (Cappellari *et al.* 2011). From the final SAURON datacubes, which were adaptively binned to a minimum signal-to-noise ratio threshold of 40, a number of quantities were derived. Gas and stellar kinematics were extracted via a pixel fitting algorithm. Emission-corrected absorption line indices in the Lick/IDS system (Trager *et al.* 1998) were derived for each bin following the procedures described in Kuntschner *et al.* (2006). Due to our limited wavelength range (4800 – 5380 Å) only the Hβ, Fe5015 and Mg b indices can be measured over the full spatial extent of the datacubes (see also McDermid *et al.* 2014).

As described in Scott *et al.* (2013), using the ATLAS^{3D} line-strength maps, we constructed radial profiles extracted from elliptical annuli for each galaxy, except for seven objects where the limited spatial extent and/or limited S/N did not warrant derivation of gradients. This leaves a total sample of 253 galaxies for this study. The ellipticity chosen was a global ellipticity measured using the weighted second moments of the surface brightness as described in Krajnović *et al.* (2011). The error on each point for the line-strengths is the rms sum of the measurement errors and the rms scatter within each

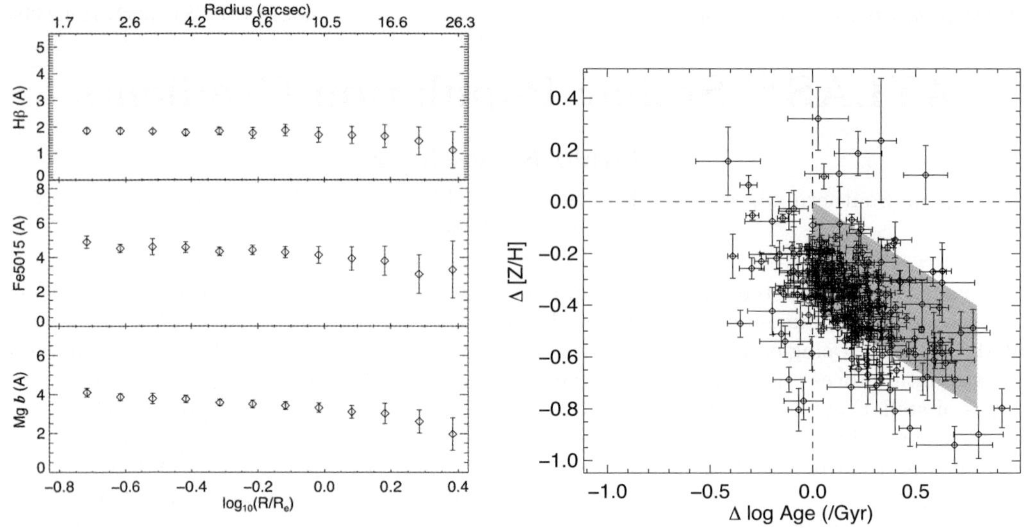

Figure 1. Left: Example of line-strength extractions in elliptical annuli as function of radius normalised to the effective radius R_e. The data shown is for NGC 2481. **Right:** Age gradients $\Delta \log \mathrm{Age}$ versus metallicity gradients $\Delta\,[Z/H]$. The grey shaded area shows the approximate region occupied by the simulations of Hopkins *et al.* (2009, their fig. 29).

annulus. Foreground stars and bad bins due to low S/N were masked in the line-strength maps before extracting the profiles. In Fig. 1 (left) we show an example of the extracted line-strength measurements as function of radius. Averaging along elliptical annuli provides high quality information on the *average* line-strengths gradients but naturally, line-strength information on more detailed spatial scales (e.g. young star forming disks or bars) is largely lost.

2. Results

On average, the galaxies in our sample show negative metallicity gradients with a robustly estimated mean gradient of $\Delta[Z/H] = -0.37 \pm 0.01(0.22)$ (a robust estimate of the 1-sigma scatter is given in brackets). The average age and $[\alpha/Fe]$ gradients are positive with values of $\Delta \log \mathrm{Age} = 0.18 \pm 0.01(0.19)$ and $\Delta[\alpha/Fe] = 0.05 \pm 0.01(0.13)$, respectively. While average negative metallicity gradients are found in essentially all studies of ETGs, results for age and $[\alpha/Fe]$ gradients do vary (e.g. Spolaor *et al.* 2010; Rawle *et al.* 2010; Koleva *et al.* 2011). In a first step to better understand the relation of stellar population gradients with other global parameters of the galaxies we explore the relation between gradients and central age and metallicity, here defined as values derived from the central $R_e/8$ aperture (McDermid *et al.* 2014). In previous studies, significant correlations were found for the metallicity gradients as well as the age gradients with central metallicity (e.g. Rawle *et al.* 2010). Furthermore, there is an anti-correlation of the age gradient with central age where younger galaxies show more positive gradients (i.e. younger centres; Kuntschner *et al.* 2010; Koleva *et al.* 2011).

We find a highly significant anti-correlation between the age gradients with the central age (Spearman rank correlation $= -0.49$ with a significance of 9.33×10^{-17}). The on-average positive age gradients can be understood straightforwardly in the hierarchical galaxy formation picture where (multiple) merging and accretion events cause the formation of young stars preferentially in the centre of galaxies (e.g. Hopkins *et al.* 2009).

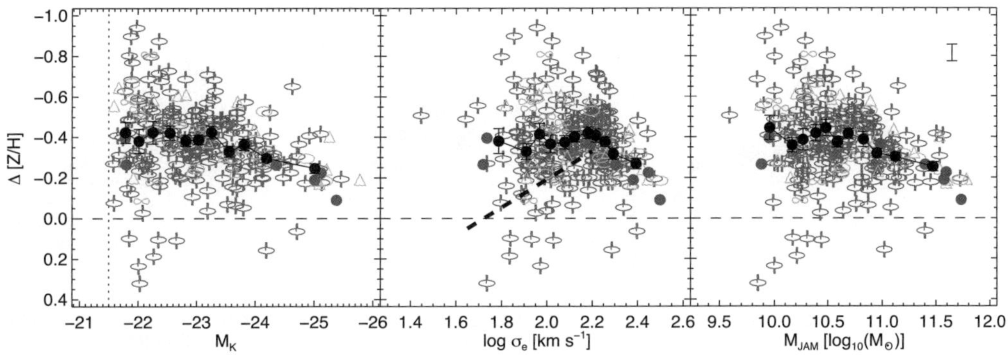

Figure 2. Metallicity gradients Δ[Z/H] are shown for 253 galaxies as function of K-band luminosity M_K, velocity dispersion within one effective radius $\log\sigma_e$, and the dynamical mass M_{JAM}. The global sample selection limited by $M_K = -21.5$ is indicated with a dotted line. Median error bars for gradients are shown in the upper right corner of the M_{JAM} panel. Red circles show galaxies with little or no rotation, green ellipses have complex velocity maps with no specific features, green triangles are KDCs, orange lemniscates have off-centre dispersion peaks indicative of superposed counter-rotating disks. Magenta spindles are the the regular rotators. The black solid symbols show robust estimates of mean values for non-overlapping bins each containing the same number of data points. Errors on those points represent robustly estimated errors on the mean. In each panel we indicate the level of zero (=no) gradient with a dashed line. The thick dashed line in the middle panel represents the tight relation of metallicity gradient with $\log\sigma_e$ found by Spolaor *et al.* (2010).

The correlation is most obvious for the "youngest" systems (with the most recent star-formation) but is still present if only galaxies with luminosity-weighted ages greater than e.g. 6 Gyr are considered (-0.37 with significance of 1.58×10^{-6}). A plausible explanation for the metallicity gradient versus central metallicity correlation also arises from the same scenario where enriched gas leads to the formation of (younger) metal rich *central* regions which steepens the overall gradients (Hopkins *et al.* 2009). If this scenario is correct, then one would expect a correlation between positive age gradients and steeper (more negative) metallicity gradients. Indeed, this is what was found previously (Rawle *et al.* 2010; Koleva *et al.* 2011) and is also true for our sample as shown in Fig. 1 (right). The Spearman rank correlation coefficient is -0.46 with a significance of 7.6×10^{-15}.

In Fig. 2 we present the relation of metallicity gradients versus three indicators of total system mass for our magnitude limited ATLAS^{3D} sample. As proxies for mass we use the integrated K-band magnitude, the velocity dispersion measured within one effective radius R_e and our own dynamical mass (M_{JAM}) estimates (Cappellari *et al.* 2013). There is a general trend of shallower metallicity gradients with increasing mass. Significant scatter is found at all masses perhaps reflecting different merger configurations and merger gas fractions, but there is an increased scatter towards the low mass end. In the $\log\sigma_e$ diagram we see a weak downturn of the relation for $\log\sigma_e < 2.2$, however, this is not as strong as what Spolaor *et al.* (2010, see thick dashed line) found. There is no obvious relation with fast or slow rotators apart from the general mass trend, where the most massive systems have the shallowest gradients and are more likely of slow rotator class.

For galaxies with extreme post-starburst-like signatures, the assumption of a *linear* stellar population gradients with $\log(R/R_e)$ can break down. Furthermore, the luminosity weighted nature and simple SSP approach of our analysis can lead to unphysical interpretations of the observed line-strength distribution for these galaxies. We therefore restrict our analysis now to galaxies with central ages of greater than 4 Gyr leaving 190

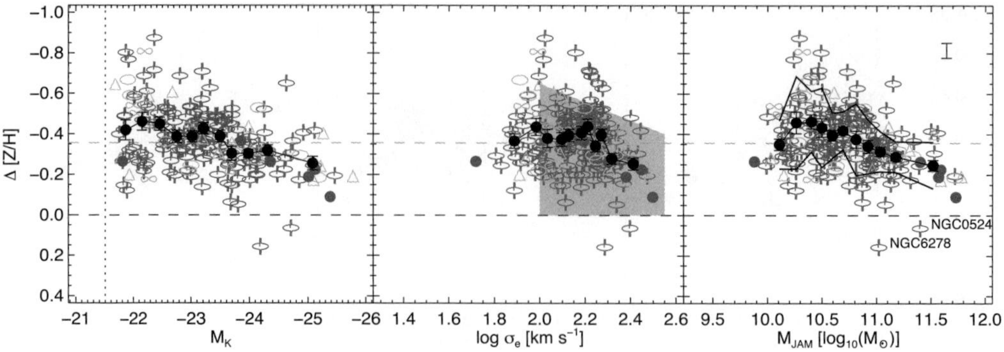

Figure 3. Metallicity gradients Δ [Z/H] are shown as function of K-band luminosity M_K, velocity dispersion within one effective radius $\log \sigma_e$, and dynamical mass M_{JAM}. The sample is restricted to 190 galaxies with central 1/8 Re ages greater than 4 Gyr. The median gradient of -0.36 dex is indicated by the dashed green line. The grey shaded area in the middle panel shows the approximate region occupied by the simulations of Hopkins *et al.* (2009, their fig. 29).

galaxies as shown in Fig. 3. The main relations persist and tighten up (i.e, less scatter), particularly the metallicity gradients versus dynamical mass relation is becoming significantly clearer (Spearman correlation of -0.33; significance 2.16×10^{-6}).

3. Conclusions

We find a significant correlation between positive age gradients (younger centres) and steeper negative metallicity gradients providing evidence in support of gas rich accretion/merging, i.e. that mergers drive gas to the centre, encouraging star formation from the enriched gas. Furthermore, we find a robustly estimated mean metallicity gradient of $\Delta[Z/H] = -0.36 \pm 0.01$ for the complete sample with a significant trend for more massive galaxies to have shallower profiles. While there is no *clear* distinction between fast and slow rotators or signs of environmental influence, we do detect a significantly larger range of [Z/H]-gradients towards low mass galaxies; this indicates that gas-rich mergers are more prevalent at lower masses.

Acknowledgements

I acknowledge generous support by the ATLAS^{3D} team.

References

Cappellari, M., Emsellem, E., Krajnović, D., *et al.*, 2011, *MNRAS*, 413, 813
Cappellari,, M., McDermid,, R. M., Alatalo, K., *et al.* 2013, *MNRAS*, 432, 1862
Hopkins, P. F., Cox, T. J., Dutta, S. N., *et al.*, 2009, *ApJS*, 181, 135
Koleva, M., Prugniel, P., De Rijcke, S., & Zeilinger, W. W., 2011, *MNRAS*, 417, 1643
Krajnović, D., Emsellem, E., Cappellari, M., *et al.*, 2011, *MNRAS*, 414, 2923
Kuntschner, H., Emsellem, E., Bacon, R., *et al.*, 2006, *MNRAS*, 369, 497
Kuntschner, H., Emsellem, E., Bacon, R., *et al.*, 2010, *MNRAS*, 408, 97
McDermid, E., *et al.*, 2014, *MNRAS*, submitted
Rawle, T. D., Smith, R. J., & Lucey, J. R., 2010, *MNRAS*, 401, 852
Scott, N., Cappellari, M., Davies, R. L., *et al.*, 2013, *MNRAS*, 432, 1894
Spolaor, M., Kobayashi, C., Forbes, D. A., Couch, W. J., & Hau, G. K. T., 2010, *MNRAS*, 408, 272
Trager, S. C., Worthey, G., Faber, S. M., Burstein, D., & Gonzalez, J. J., 1998, *ApJS*, 116, 1

Galaxy Masses as Constraints of Formation Models
Proceedings IAU Symposium No. 311, 2014
M. Cappellari & S. Courteau, eds.

doi:10.1017/S1743921315003397

Late stages of stellar evolution in population models.

Claudia Maraston

Institute of Cosmology and Gravitation, University of Portsmouth, Dennis Sciama Building, Burnaby Road, Portsmouth PO1 3FX. email: claudia.maraston@port.ac.uk

Abstract. My contribution to Roger's celebration symposium focuses on the treatment of late stellar evolutionary phases in stellar population models, reviewing the state of art and discussing some very recent developments, ranging from local stellar clusters up to distant galaxies at high redshift. I shall focus in particular on the Thermally Pulsating Asymptotic Giant Branch, about which a vivid discussion has been ongoing since a few years. I shall present renewed evidence in favour of a sizable contribution from this phase for matching the observed spectral energy distribution of distant massive galaxies. I shall also discuss the possible reasons why such a conclusion has been controversial in the recent literature. Stellar population models are the magic tool to shape the physics of galaxies out of their observed light, and enter virtually all papers presented at this symposium. In a collective effort to properly treat all relevant aspects of the modelling, we split the discussion into six contributions given by experts in the field, as our present to Roger and his outstanding career.

Keywords. stars: evolution - galaxies: evolution

---

1. Introduction

Late stages of stellar evolution refer to the period of (relatively) rapid burning of Hydrogen and/or Helium in shells outside the core, following the exhaustion of the main stellar fuel, the central Hydrogen burning. The occurrence and duration of such phases depends on the initial stellar mass. High-mass stars ($M > 8\ M_{\odot}$) go rapidly into central He-burning and after a short phase of Red or Blue supergiants, end as SN-II. In intermediate-mass stars ($2 < M < 8\ M_{\odot}$) the most relevant late evolutionary phase is the Asytmptotic Giant Branch (AGB), while low-mass stars ($M < 2\ M_{\odot}$) mark the start of prolongued Red Giant Branch (RGB) and Horizontal Branch (HB). Fig. 1 gives a snapshot of late evolutionary phases on the HR diagram of an old population.†

Late evolutionary phases are featured by short ($t < 1$ Gyr) lifetimes, a high luminosity, temperatures ranging to extreme values (the tip of the RGB and the AGB extend to very cold temperatures, $T_{\rm eff} < 4000\ K$ and hot HB runs to hot values of temperatures, $T_{\rm eff} >> 10000\ K$). The extreme regimes coupled with mass-loss, rotation and envelope instabilities make the theoretical modelling complicated. Mass-loss in particular hampers a clean prediction of the full evolution over key late phases such as the pulsating major part of the AGB, so-called Thermally-Pulsating Asymptotic Giant Branch (TP-AGB), and the Horizontal Branch. In order to encompass these difficulties and help calculating models which could be applied to galaxies with some confidence, Renzini & Fusi-Pecci (1988) suggested to calibrate the unknown of stellar evolution on the simplest stellar systems known in order to obtain empirical estimates of a few relevant parameters, such as the efficiency of mass-loss, the fuel burnt, etc. which could then be used to

† There are other late phases, e.g. the UV-upturn, which are not treated here for reasons of space.

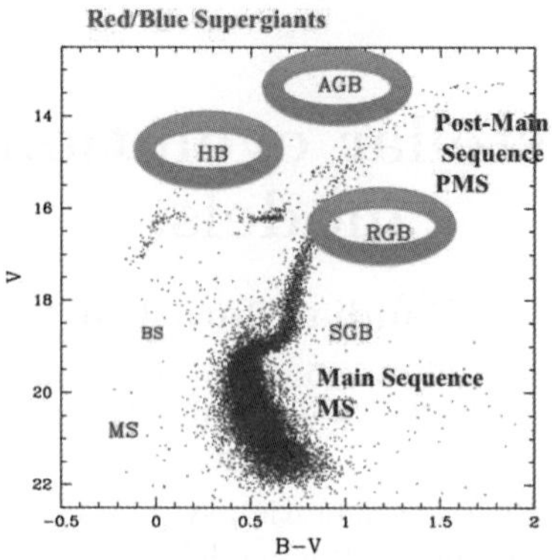

Figure 1. Displacement of late stellar evolutionary phases, RGB, HB, AGB (circled in orange) and red/blue supergiants, labelled on the upper axis, onto the HR diagram of an old population. Adapted from Maraston (2005).

constrain theoretical stellar models. A similar approach was then extended to stellar population (i.e. integrated) models by using whole star clusters as proto-typical single burst (SSP) populations (for the TP-AGB Maraston 1998; Maraston & Thomas 2000, for the HB morphology). Even if nowadays the actual nature of most star clusters as single generations is questioned as often multiple Main Sequences are detected in Milky Way and Magellanic Clouds objects (e.g. Piotto *et al.* 2007), star clusters are still simpler than galaxies.

They do have a disadvantage though, which is their relatively low number of stars ($\sim 10^{4\div6}$) which makes the sampling of the short late stelar stages hard against stochastical magnitude fluctuations. Massive galaxies hosting nearly coeval stellar populations are immune from fluctuations. This fact was the pillar for using passive galaxies at high redshift as a showcase for the testing of various assumptions regarding the duration and fuel consumption of the TP-AGB (Maraston *et al.* 2006), the shortest ($t \sim 10^6$ yr) and most luminous late phase of stellar evolution.

The energetic and lifetime we include in population models directly impact on the physical properties that are derived or predicted for galaxies, hence on our understanding of galaxy evolution. In the following I shall briefly review the most important facts of each late stellar phases, allowing more space to the highly debated TP-AGB and new results first presented at this symposium. I shall then provide a brief summary.

2. Massive stars: Red or Blue supergiants?

The fate of a massive star as a red or blue supergiant depends on assumptions for convective overshooting, rotation and mass-loss (see e.g. Chiosi & Maeder 1986). Various stellar models include this late stage of massive stars with a different typical temperature, which then reflects itself in a different integrated spectrum of stellar populations. Fig. 2 (left-hand panel) compares models by Maraston (2005, hereafter M05) (red) and Bruzual & Charlot (2003, hereafter BC03) (blue) which are based on Geneva and Padova stellar tracks, respectively, at ages of 1 and 10 Myr. The effect of the temperature of the supergiant phase is evident at 10 Myr models, where the M05 model has more near-IR flux than the BC03 one, because the phase is spent at cooler temperatures. How to disentangle the correct temperature? The right-hand panel is taken from Marigo *et al.* (2008), which uses the $V - K$ integrated colour vs age as a diagnostic to compare stellar population models (lines) with Magellanic Clouds star clusters (symbols). The youngest

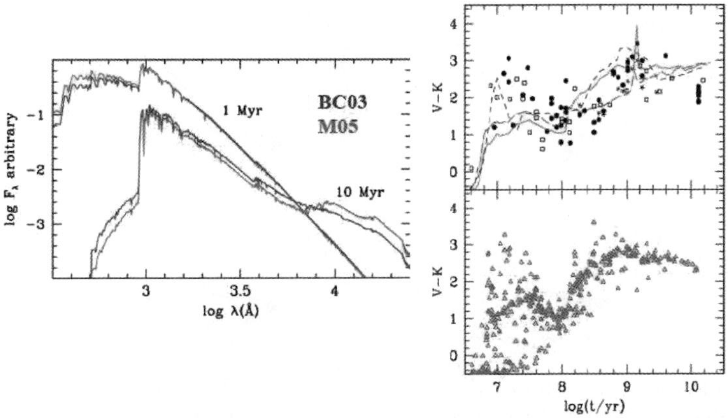

Figure 2. The left-hand panel shows young SSP models by M05 (red) and BC03 (blue), which are based on different stellar evolutionary tracks. The right-hand panel from Marigo *et al.* (2008) shows how SSP models compare to the $V - K$ colours of Magellanic Clouds GCs. M05 are the blue lines.

clusters reach very red colours, which are better consistent with a red supergiant phase as in the M05 models (blue lines). Marigo *et al.* (2008) also shows in the lower panel the effect of stochastical fluctuations on their stellar population models based on Padova tracks.

3. The temperature of the Red Giant Branch.

Maraston (2005, Figure 9) shows how different stellar models predict different typical temperatures for the RGB, the tracks by the Padova database being the coolest and those from the Teramo's one the warmest. Fig. 27 in the same paper shows the RGB temperature effect on the integrated spectrum of a stellar population model. The coolest tracks make the integrated SED redder which, in terms of galaxy properties, implies the derivation of a lower total metallicity or a younger age. The RGB flux vs $T_{\rm eff}$ also affects the derivation of the stellar Initial Mass Function (IMF) from near-IR indices (see C. Conroy; R. Smith, this volume). The RGB track depends on several input physics, notably on the mixing-length which is one of the parameters to be calibrated with observations (see discussion in M05). Interesting plots are shown in McDonald *et al.* (2011) who use various stellar models to derive the properties of stars in the metal-rich MW GC 47 Tuc (Fig. 3). It can be seen that Padova tracks are indeed too cold at the nominal cluster metallicity (left-hand panel), whereas those by the Darmouth† database fit better (right-hand panel). However, the same tracks including an α/Fe ratio as measured in 47 Tuc (lower right-hand panel) perform slightly worse than the solar-scaled ones! Progress is needed to correctly predict the effect of non-solar chemical mixtures on stellar evolution and model atmospheres (see D. Thomas and R. Schiavon, this volume).

4. The morphology of the Horizontal Branch.

It is well known that the late evolutionary phase of HB manifests itself in a variety of 'morphologies', thought to be mostly (but not only) related to metallicity (see references in Maraston & Thomas (2000)). Stellar population models calculated for various HB

† Due to space constraints we cannot provide full citations and refer the reader to the original papers.

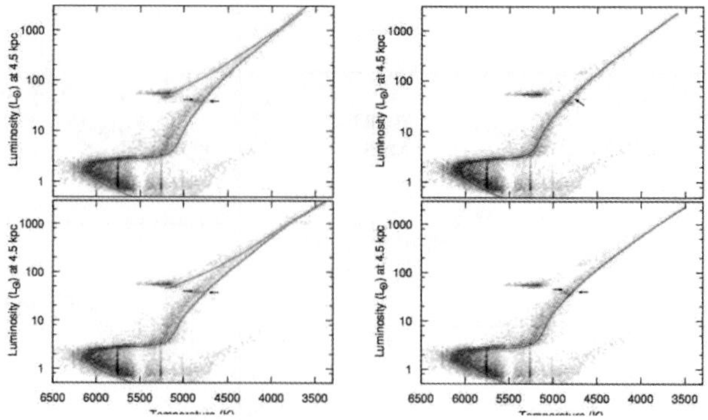

Figure 3. Comparison of theoretical isochrones, from Padova and Darmouth (left-hand and right-hand panels, respectively) to the observed CMD of the metal-rich GC 47 Tuc. From McDonald *et al.* (2011).

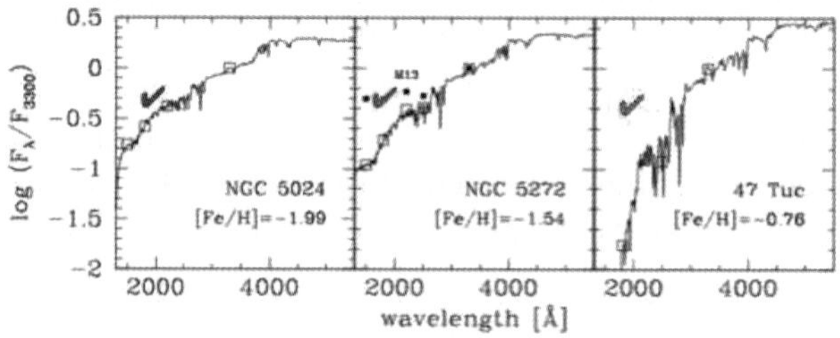

Figure 4. Theoretical SED of stellar population models with the same age (14 Gyr) and various [Z/H] matching the variety of HB morphologies observed in MW GCs, blue, intermediate and red from left to right. From Maraston & Thomas (2000).

morphologies (as in M05) have a larger degree of flexibility to experiment with observations. In the M05 models, the tunable parameter is the mass-loss along the RGB, which allows to obtain hotter tracks without varying the evolutionary mass (i.e. the age) of the model. Fig. 4 plots data for three MW GCs having different metallicities, a range of HB morphologies, from blue to red (from left to right) and a similar age. Stellar population models are best-fitted to them, in which the HB morphology has been adapted to the observed case, by playing with the mass-loss efficiency.

5. The infamous TP-AGB phase and high-redshift galaxies.

The TP-AGB phase and its effects on stellar population models and galaxies has gained the honour of the press several times since the works of Maraston (1998, 2005); Maraston *et al.* (2006). Several reviews are available (e.g. Maraston 2013), the content of which will not be repeated here. Briefly, Maraston (1998, 2005) used Magellanic Clouds GCs to calibrate the energetic of the TP-AGB and use the observed colours of the same clusters to fix the flux of model SEDs. **?** (and others) find the M05 models to match well the full SEDs of galaxies at high-z, in particular those in mostly passive evolution. Recently, Conroy & Gunn (2010) have questioned the above mentioned calibration, based on newer MC GCs data and especially on a new age scale for the observed clusters, concluding that M05 models are too red, in other words contain too much TP-AGB. On a similar path, Kriek *et al.* (2010) concluded that BC03 models fit observed galaxy SEDs much better than M05 models and indicate in the high TP-AGB of the M05 models the cause of the misfit.

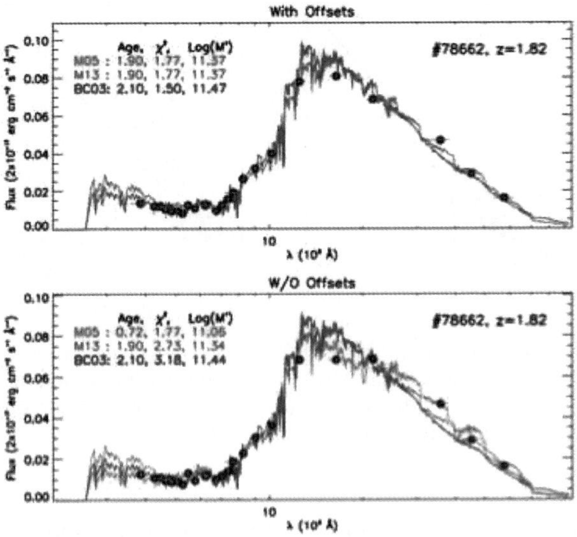

Figure 5. Effect of model-dependent photometric offsets on SED fits. Upper panel: data with model-dependant offsets favour BC03 models (blue line). Bottom panel: the removal of the offsets changes the observed SED and favour M05 (red line). Adapted from Capozzi *et al.* (2015).

The Kriek *et al.* (2010) sample contain $\sim$ 66 galaxies with photometric redshift, pre-selected and stack as to match a post-starburst criterium defined via rest-frame colours. Their approach is quite dissimilar to that of Maraston *et al.* (2006), hence Capozzi *et al.* (2015) re-examine our earlier results, who strictly used galaxies with spectroscopic red-shifts only, by re-analysing the same objects with new photometry available in the mean-time. Moreover, the sample of spectroscopic galaxies was augmented with $\sim$ 50 more objects from the COSMOS survey pre-selected to be in nearly passive evolution as those of Maraston *et al.* (2006). It should be noticed that this selection is applied initially, but then the data are fit with models calculated with a large variety of star formation histories and reddening, i.e. the galaxies are left free to move away from this initial pre-selection, wish they wanted to.

Capozzi *et al.* (2015) fully confirm the results of Maraston *et al.* (2006), namely that the strong TP-AGB M05 models fit the data better than BC03 models, and the inclusion of reddening as an additional free parameter makes the fits with both models comparable. These results are ad odd with Kriek *et al.* (2010). Capozzi *et al.* (2015) discuss another interesting effect. It is found that often raw photometric data are applied offsets to improve the performance of photometric redshifts. These offsets are calculated using stellar population models. It should then go without saying that 'model-shaped' data should not be further used to perform comparisons with models, as obviously the models used to obtain the offsets would appear favoured. The case is shown in Fig. 5 and 6. It remains to be assessed whether the use of BC03-based off-setted photometry is in part responsible for the discrepant results just discussed.

Finally, the new version of M05-type models with reduced TP-AGB fuel as re-calibrated after Conroy & Gunn (2010) (shown in Noell *et al.* 2013) are not found to match the high-z galaxies better than M05.

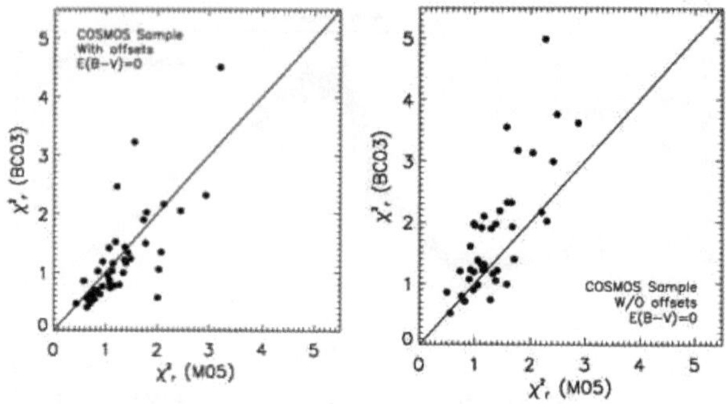

Figure 6. The effect of model-dependent photometric offsets to galaxy data on the goodness of SED fits of models to the same data. Model-dependant offsets favour the models on which the offsets were calculated (BC03, left-hand panel). The removal of the offset changes the trend and now M05 is favoured. Adapted from Capozzi *et al.* (2015).

6. Outlook

A proper description of late evolutionary stages in stellar population models requires the calibration of free parameters with data, and stellar clusters have been regarded as the ideal objects in this game. While the procedure still seems to hold for most phases, an interesting discrepancy emerges between star clusters and massive galaxies in case of the TP-AGB phase, where galaxies require a large effect from this stellar phases.

Acknowledgements

Thanks to Michele and Stephane for organising an outstanding conference, and to Roger for being such a brilliant scientist and a great friend.

References

Bruzual, G. & Charlot, S. 2003, *MNRAS*, 344, 1000
Capozzi, D. *et al.*, 2015, *MNRAS*, *submitted*
Chiosi, C. & Maeder, A. 1986, *ARA&A*, 24, 329
Conroy, C. & Gunn, J. 2010, *ApJ*, 712, 883
Kriek, M., *et al.* 2010, *ApJ*, 722, L64
Maraston, C., 1998, *MNRAS*, 300, 872
Maraston, C., 2005, *MNRAS*, 362, 799
Maraston, C., 2013, in "*The Intriguing Life of Massive Galaxies, Proceedings of the International Astronomical Union*, IAU Symposium, Volume 295, pp. 272-281
Marigo, P., *et al.* 2008, å, 482, 883
Maraston, C., Thomas, D. 2000, *ApJ*, 541, 126
Maraston, C., Daddi, E., & Renzini, A. 2006, *ApJ*, 652, 85
McDonald, A., *et al.* 2011, *ApJS*, 193, 23
Noel, N., *et al.* 2013, *ApJ*, 772, 58
Piotto, G., *et al.* 2007, *ApJ*, 661, L53
Renzini, A. & Fusi-Pecci, F. 1988, *ARA&A*, 26, 199

Galaxy Masses as Constraints of Formation Models
Proceedings IAU Symposium No. 311, 2014
M. Cappellari & S. Courteau, eds.

doi:10.1017/S1743921315003403

Element abundance ratios in stellar population modelling

Daniel Thomas

Institute of Cosmology and Gravitation, University of Portsmouth, Dennis Sciama Building, Burnaby Road, Portsmouth, PO1 3FX, UK. email: daniel.thomas@port.ac.uk
SEPnet, South East Physics Network, (www.sepnet.ac.uk)

Abstract. I review the implementation of the effects from varying chemical element abundance ratios in stellar population modelling, focusing on α- and Fe-peak elements. A brief overview of the development of such models over the past 30 years is provided, starting with early work on the identification of relevant absorption features in the spectra of early-type galaxies in the 1980s leading to the most recent developments of the past years. Recent highlights include the adoption of new flux calibrated libraries, the inclusion of a wide range of chemical elements, the calculation of error estimates on the model, and the consideration of element variation effects on full spectra. The calibration of such models with globular clusters and some key results on the element ratios measured in early-type galaxies are presented.

Keywords. galaxies: elliptical and lenticular, cD - galaxies: evolution - galaxies: formation

1. Introduction

The spectra of galaxies and globular clusters carry a wealth of information about gas and stellar population properties. Emission lines are used to derive gas kinematics, star formation activity and black hole accretion (e.g. Kauffmann *et al.* 2003; Tremonti *et al.* 2004; Sarzi *et al.* 2006; Schawinski *et al.* 2007; Thomas *et al.* 2013), the absorption line and stellar continuum component of the spectrum, instead, discloses stellar population properties that act as fossil record and can be used to derive ages, formation epochs, star formation histories, and element abundances (e.g. Trager *et al.* 2000a; Kuntschner 2000; Thomas *et al.* 2005; Nelan *et al.* 2005; Thomas *et al.* 2010). The absorption features in a spectrum are particularly useful for several reasons. They can easily be measured and calibrated to a common system (Burstein *et al.* 1984; Faber *et al.* 1985), they are largely insensitive to dust attenuation (MacArthur 2005), and they allow to dissect 'metallicity' into individual element abundances (Greggio 1997; Tantalo *et al.* 1998; Trager *et al.* 2000b; Thomas *et al.* 2003a) that in turn set valuable constraints on the chemical enrichment history (Thomas *et al.* 1999). The inclusion of the these effects in stellar modelling has produced a major step forward in the field of stellar population modelling and has led to key progress in the interpretation of galaxy data.

2. Brief history of element ratio modelling

In the 1980s the Lick group have defined optical absorption-line indices (Burstein *et al.* 1984; Faber *et al.* 1985; Gorgas *et al.* 1993) that are by far the most commonly used in absorption-line analyses of old stellar populations. The bandpasses of Lick indices are relatively large with widths up to 50 Å and two windows blue- and redward of the band-pass defining pseudo-continua. They have been designed for massive galaxies with significant line broadening through random motions of the stars. The advantage is that a

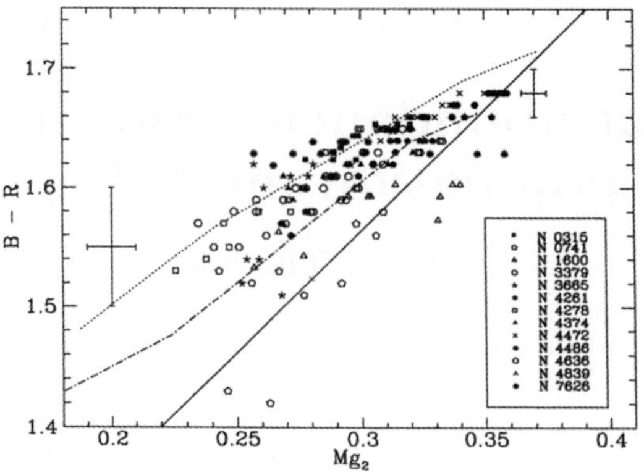

Figure 1. Measurements of Mg vs Fe absorption line indices for early-type galaxies confronted with stellar population models. From Davies *et al.*(1993).

wide bandpass increases the signal-to-noise ratio, and measurements are relatively robust. The downside is that they are an agglomeration of a large number of absorption lines from various chemical elements, so that their use for the derivation of individual element abundances is nontrivial.

While the 1990s have seen the advent of a number of stellar population models produced for the aim of analysing galaxy spectra (Buzzoni 1989; Bruzual & Charlot 1993; Vazdekis *et al.* 1996; Worthey 1994; Maraston 1998), it was also recognised that Mg and Fe absorption features in early-type galaxies could not be reproduced by these models (Peletier 1989; Worthey *et al.* 1992; Davies *et al.* 1993; Carollo & Danziger 1994; Bender & Paquet 1995; Greggio 1997) as shown in Fig. 1. Tripicco & Bell (1995) made a critical step forward and determined the sensitivity of Lick absorption-line indices to individual element abundance variations through model atmosphere calculations.

The first stellar population models with variable abundance element ratios using these calculations were then generated by Thomas *et al.* (2003a) based on a method developed by Tantalo *et al.* (1998) and Trager *et al.* (2000b). A key in the development of these models was the empirical calibration with globular cluster data by Maraston *et al.* (2003). The models have subsequently been updated with new model atmosphere calculations by Korn *et al.* (2005, see also Houdashelt *et al.*2005) for non-solar metallicities and the higher-order Balmer line indices. The latter turned out to be key for the interpretation of bluer Balmer absorption strengths in galaxies. Because of the presence of significant Fe absorption in the pseudo continua of the Balmer indices Hγ and Hδ, both features are highly sensitive to α/Fe ratios leading to relatively strong index strengths in old, metal-rich stellar populations (Thomas *et al.* 2004). The inclusion of this effect is critical for the correct derivation of galaxy ages from these features (Thomas & Davies 2006).

More recently, a number of other element abundance sensitive models have been produced increasing the number of chemical elements considered, implementing other stellar libraries, and extending the approach to full spectral response (Annibali *et al.* 2007; Schiavon 2007; Coelho *et al.* 2007; Lee *et al.* 2009; Thomas *et al.* 2011; Worthey *et al.* 2011; Conroy *et al.* 2014).

3. Recent progress

In the following I will discuss in more detail some of the recent developments.

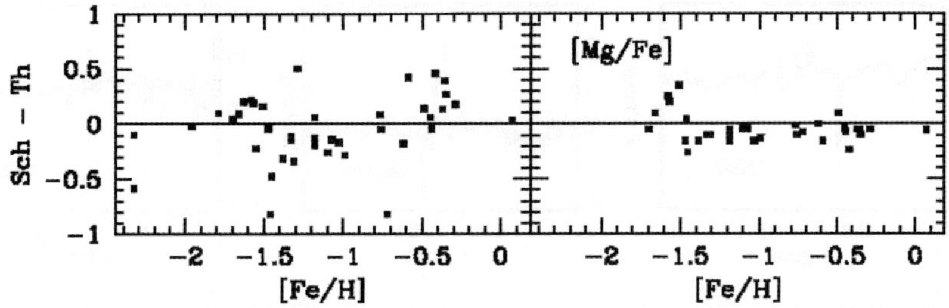

Figure 2. Globular cluster element abundances [Fe/H] and [Mg/Fe] from Thomas *et al.* (2011) and Schiavon (2007) models. R. Schiavon (private communication).

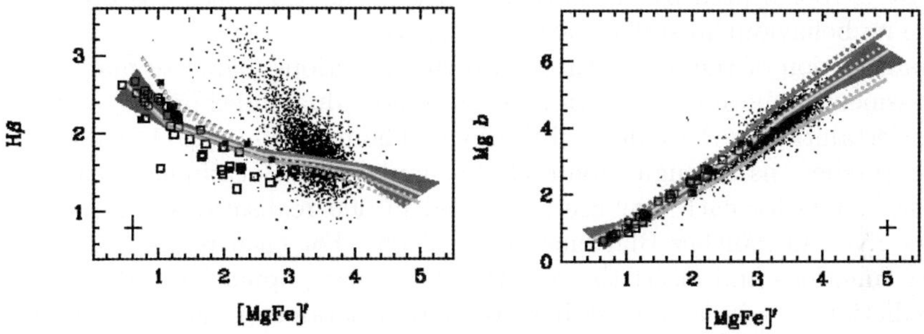

Figure 3. Calibration of the Hβ and Mgb indices with galactic globular clusters (squares). Models with an age of 13 Gyr and various metallicities and α/Fe ratios are shown. The grey shaded area along the model with α/Fe] = 0.3 dex (yellow line) indicates the 1-σ error of the model prediction. The small black dots are SDSS early-type galaxies from Thomas *et al.* (2010). From Thomas *et al.* (2011).

3.1. *New stellar libraries*

The advent of new flux-calibrated libraries has made it possible to produce models that are now flux calibrated and at higher spectral resolution, and in particular not tied anymore to the Lick/IDS system. This is essential for the interpretation of galaxy spectra where calibration stars are not available, such as large galaxy redshift surveys or other high-redshift observations. Johansson *et al.* (2010) produced new empirical fitting functions based on the MILES library (Sánchez-Blázquez *et al.* 2006) that relate absorption line strength with stellar parameters, that are adopted in the Thomas *et al.* (2011) models. These models can directly be applied to SDSS data with minimal correction for spectral resolution (Beifiori *et al.* 2011).

The models have been calibrated with galactic globular clusters (Puzia *et al.* 2002; Schiavon *et al.* 2005) in Thomas *et al.* (2011). Ages and total metallicities and some element ratios are well reproduced, while some element ratios remain uncertain. It is encouraging, though, that different models give consistent result for some element ratios as shown in Fig. 2 presenting the comparison of globular cluster element abundances [Fe/H] and [Mg/Fe] from Thomas *et al.* (2011) and Schiavon (2007) models.

3.2. *Inclusion of model errors*

A particular feature of the Thomas *et al.* (2011) models is that model errors are calculated for the first time. This is possible thanks to the use of absorption line strengths in combination with fitting functions. The major strength of fitting functions lies in the fact that they allow for interpolation between well-populated regions of stellar parameter space which increases the accuracy of the model in stellar parameter space that is only sparsely sampled by empirical stellar libraries. Moreover, each absorption index or spec-

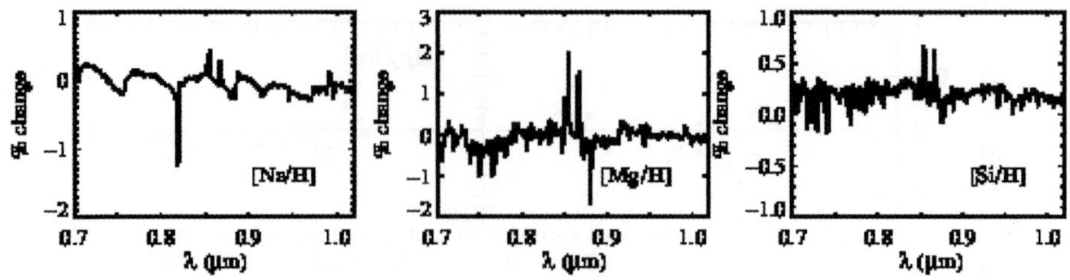

Figure 4. An example of the wavelength dependent response to the variation of the elements Na, Mg, and Si. From Conroy *et al.* (2014).

tral feature is represented by an individual fitting function, which is optimised to best reproduce its behaviour in stellar parameter space.

The construction of the model through fitting functions allows to make a straightforward assessment of the statistical errors on each individual index prediction that is caused by the uncertainties in the index calibrations. Thomas *et al.* (2011) calculate errors in the model predictions through Monte Carlo simulations. Both index measurements and stellar parameters for each star are perturbed in 600 realisations per simulation using the errors given in Sánchez-Blázquez *et al.* (2006). For each realisation they re-derive the fitting functions and insert those in the the stellar population code. This yields 600 index predictions per index and stellar population parameter age, metallicity, and α/Fe ratio. They then fit a Gaussian to the distribution of index strengths and derive the 1-σ error. Note that these errors do not include systematic effects such as a change of stellar evolutionary track, for instance.

Fig. 3 illustrates the effect. Model errors are generally very small and well below the observational errors for globular clusters or galaxies around solar metallicity. Model errors rise considerably and become comparable to the typical observational error toward the highest and lowest metallicities. This behaviour ought to be expected and is a direct consequence of the empirical stellar library, in which the stellar parameter space is inevitably sampled worse at the ends of the distribution. As a consequence, model errors turn out to be quite large at the edges. An increase of sampling density in stellar libraries would therefore be highly beneficial in these sections of the parameter space.

3.3. *Full spectral response*

An alternative to the approach of Thomas *et al.* (2003a) has been presented in the recent literature. Conroy *et al.* (2014) follow the method of Thomas *et al.* (2003a) in that they use response functions from model atmosphere calculations to assess the relative effect of element abundance variations, while the basic stellar population model is footed on empirical stellar libraries. Rather than using response functions for key absorption features, though, they calculate wavelength-dependent spectral response functions and apply those to the full model spectrum. This approach is very powerful but somewhat more affected by uncertainties in line lists and continuum uncertainties, hence may be considered well complementary to the use of absorption index features. An example of the wavelength dependent response to the variation of the elements Na, Mg, and Si is shown in Fig. 4.

3.4. *Galaxy data*

Element ratio sensitive stellar population models provide the opportunity to derive the abundance ratios for a variety of chemical elements of unresolved stellar populations. Graves & Schiavon (2008) and Johansson *et al.* (2012) developed codes that fit iteratively a set of absorption line indices resulting in the derivation of the abundance ratios of the

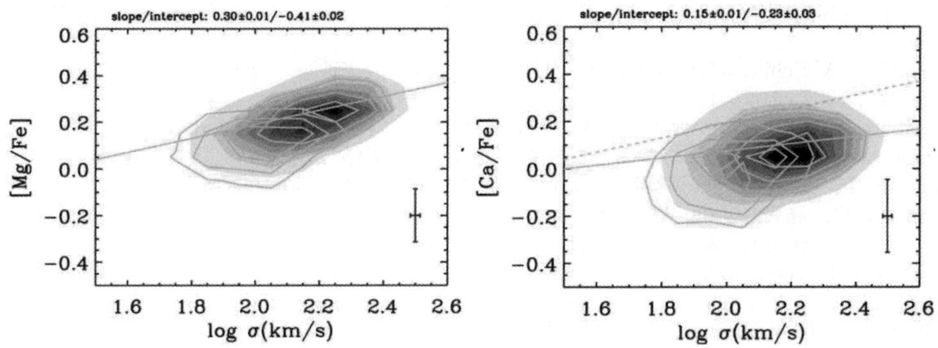

Figure 5. Element ratios [Mg/Fe] and [Ca/Fe] as a functions of velocity dispersion in early-type galaxies. The solid line is a linear fit, the dotted line shows the correlation for [Mg/Fe]. From Johansson *et al.* (2012).

elements best accessible through strong absorption features including C, N, Mg, Ca, Ti, and Fe. The abundance of oxygen cannot be derived directly from absorption features, but can be estimated through its role as dominating element among the light α-elements (Johansson *et al.* 2012). Conroy *et al.* (2014) recently extended this procedure to full spectral fits adding the element Si.

A key result seen in all these studies consistently is that the light-element to Fe-peak abundance ratios [C/Fe], [N/Fe], [O/Fe], [Mg/Fe], and [Si/Fe] increase with galaxy velocity dispersion and mass (Smith *et al.* 2009; Johansson *et al.* 2012; Conroy *et al.* 2014), leading to significantly super-solar ratios in massive galaxies. The heavier elements Ca and Ti deviate from this trend. [Ca/Fe] is close to solar across all galaxy masses and may only increase slightly (if at all) with galaxy mass (see Fig. 5). This has also been interpreted as Calcium under-abundance in massive galaxies (Thomas *et al.* 2003b). The best explanation is that the relevance of Type Ia enrichment increases with element mass, which implies that Ca and Ti are suppressed because they get partially enriched by the delayed Type Ia supernovae. However, Ti, even though being heavier than Ca, appears to show somewhat higher enhancement than Ca in massive galaxies, which seems contrived within this picture. The optical spectrum used in these studies only is moderately sensitive to Ti abundance, though, and further work on this, potentially including the near-IR part of the spectrum, will be very valuable.

References

Annibali, F., Bressan, A., Rampazzo, R., Zeilinger, W. W., & Danese, L., 2007, *A&A*, 463, 455

Beifiori, A., Maraston, C., Thomas, D., & Johansson, J., 2011, *A&A*, 531, 109

Bender, R. & Paquet, A., 1995, in van der Kruit P. C., Gillmore G., eds, Stellar Populations IAU Symposium 164. Kluwer Academic Publishers, Dordrecht, p. 259

Bruzual, A. G. & Charlot S., 1993, *ApJ*, 405, 538

Burstein, D., Faber, S. M., Gaskell, C. M., & Krumm, N., 1984, *ApJ*, 287, 586

Buzzoni, A., 1989, *ApJS*, 71, 817

Carollo, C. M. & Danziger, I. J., 1994, *MNRAS*, 270, 523

Coelho, P., Bruzual, G., Charlot, S., Weiss, A., Barbuy, B., & Ferguson, J. W., 2007, *MNRAS*, 382, 498

Conroy, C., Graves, G. J., & van Dokkum, P. G., 2014, *ApJ*, 780, 33

Davies, R. L., Sadler, E. M., & Peletier, R. F., 1993, *MNRAS*, 262, 650

Faber, S. M., Friel, E. D., Burstein, D., & Gaskell, D. M., 1985, *ApJS*, 57, 711

Gorgas, J., Faber, S. M., Burstein, D., González, J. J., Courteau, S., & Prosser, C., 1993, *ApJS*, 86, 153
Graves, G. J. & Schiavon, R. P., 2008, *ApJS*, 177, 446
Greggio, L., 1997, *MNRAS*, 285, 151
Houdashelt, M. L., Trager, S. C., & Worthey, G., 2005, Highlights of Astronomy, 13, 585
Johansson, J., Thomas, D., & Maraston, C., 2010, *MNRAS*, 406, 165
Johansson, J., Thomas, D., & Maraston, C., 2012, *MNRAS*, 421, 1908
Kauffmann, G., *et al.*, 2003, *MNRAS*, 341, 33
Korn, A., Maraston, C., & Thomas, D., 2005, *A&A*, 438, 685
Kuntschner, H., 2000, *MNRAS*, 315, 184
Lee, H., Worthey, G., Dotter, A., Chaboyer, B., Jevremović, D., Baron, E., Briley, M. M., Ferguson, J. W., Coelho, P., & Trager, S. C., 2009, *ApJ*, 694, 902
MacArthur, L. A., 2005, *ApJ*, 623, 795
Maraston, C., 1998, *MNRAS*, 300, 872
Maraston, C., Greggio, L., Renzini, A., Ortolani, S., Saglia, R. P., Puzia, T., & Kissler-Patig, M., 2003, *A&A*, 400, 823
Nelan, J. E., *et al.*, 2005, *ApJ*, 632, 137
Peletier, R., 1989, Phd thesis, Rijksuniversiteit Groningen
Puzia, T., Saglia, R. P., Kissler-Patig, M., Maraston, C., Greggio, L., Renzini, A., & Ortolani, S., 2002, *A&A*, 395, 45
Sánchez-Blázquez, P., Peletier, R. F., Jiménez-Vicente, J., Cardiel, N., Cenarro, A. J., Falcón-Barroso, J., Gorgas, J., Selam, S., & Vazdekis, A., 2006, *MNRAS*, 371, 703
Sarzi, M., Falcón-Barroso, J., Davies, R. L., Bacon, R., Bureau, M., Cappellari, M., de Zeeuw, P. T., Emsellem, E., Fathi, K., Krajnović, D., Kuntschner, H., McDermid, R. M., & Peletier, R. F., 2006, *MNRAS*, 366, 1151
Schawinski, K., Thomas, D., Sarzi, M., Maraston, C., Kaviraj, S., Joo, S.-J., Yi, S. K., & Silk, J., 2007, *MNRAS*, 382, 1415
Schiavon, R. P., 2007, *ApJS*, 171, 146
Schiavon, R. P., Rose, J. A., Courteau, S., & MacArthur, L. A., 2005, *ApJS*, 160, 163
Smith, R. J., Lucey, J. R., Hudson, M. J., & Bridges, T. J., 2009, *MNRAS*, 398, 119
Tantalo, R., Chiosi, C., & Bressan, A., 1998, *A&A*, 333, 419
Thomas, D. & Davies, R. L., 2006, *MNRAS*, 366, 510
Thomas, D., Greggio, L., & Bender, R., 1999, *MNRAS*, 302, 537
Thomas, D., Johansson, J., & Maraston, C., 2011, *MNRAS*, 412, 2199
Thomas, D., Maraston, C., & Bender, R., 2003a, *MNRAS*, 339, 897
Thomas, D., Maraston, C., & Bender, R., 2003b, *MNRAS*, 343, 279
Thomas, D., Maraston, C., Bender, R., & Mendes de Oliveira, C., 2005, *ApJ*, 621, 673
Thomas, D., Maraston, C., & Johansson, J., 2011, *MNRAS*, 412, 2183
Thomas, D., Maraston, C., & Korn, A., 2004, *MNRAS*, 351, L19
Thomas, D., Maraston, C., Schawinski, K., Sarzi, M., & Silk, J., 2010, *MNRAS*, 404, 1775
Thomas, D., Steele, O., Maraston, C., Johansson, J., *et al.*, 2013, *MNRAS*, 431, 1383
Trager, S. C., Faber, S. M., Worthey, G., & González, J. J., 2000a, *AJ*, 120, 165
Trager, S. C., Faber, S. M., Worthey, G., & González, J. J., 2000b, *AJ*, 119, 164
Tremonti, C. A., Heckman, T. M., Kauffmann, G., *et al.*, 2004, *ApJ*, 613, 898
Tripicco, M. J. & Bell, R. A., 1995, *AJ*, 110, 3035
Vazdekis, A., Casuso, E., Peletier, R. F., & Beckmann, J. E., 1996, *ApJS*, 106, 307
Worthey, G., 1994, *ApJS*, 95, 107
Worthey, G., Faber, S. M., & González, J. J., 1992, *ApJ*, 398, 69
Worthey, G., Ingermann, B. A., & Serven, J., 2011, *ApJ*, 729, 148

Galaxy Masses as Constraints of Formation Models
Proceedings IAU Symposium No. 311, 2014
M. Cappellari & S. Courteau, eds.

doi:10.1017/S1743921315003415

NaD excess objects and its implications on initial mass function

Sukyoung K. Yi[1] **and Hyunjin Jeong**[2]

[1]Yonsei University, Astronomy Department, Seoul 120-749, Republic of Korea. email: yi@yonsei.ac.kr

[2]Korea Astronomy and Space Science Institute Daejeon 305-348, Republic of Korea. email: hyunjin@kasi.re.kr

Abstract. van Dokkum and Conroy (2010) have heralded the possible implication of NaI 8200A lines on extra bottom-heavy IMF of bright elliptical galaxies. A flurry of papers have followed it and revisited the half a century old issue all over again from various angles. For a sanity check, we investigate on the issue this time using the better measured NaD 5900A line instead. We found from the SDSS database that a large ($\sim$ 10%) fraction of galaxies (both early and late types) exhibit strong NaD lines and that the same bottom-heavy models of van Dokkum and Conroy that reproduced the strong NaI 8200 lines of elliptical galaxies are not capable of reproducing NaD line strengths at all. The mystery prevails. This article is mainly a review of our recent paper (Jeong *et al.* 2013).

Keywords. galaxies: elliptical and lenticular, cD - galaxies: evolution - galaxies: formation

1. Introduction

The fact that the apparently boring spectra of elliptical galaxies have many hints to their formation history is evident and it was properly highlighted by van Dokkum and Conroy through their series of papers (van Dokkum and Conroy 2010; 2011; 2012). Na lines (mostly NaD 5900) have been particularly of interest ever since their strengths were noticed to be much stronger in bright elliptical galaxies than expected for "normal?? stellar populations (O?Connell 1976; Peterson 1976; Hardy and Couture 1988; Worthey 1988).

van Dokkum and Conroy in their series of papers reported that the NaI 8200 line strengths in the near infrared are also very strong in bright ellipticals and that such strong line strengths can be explained if the bulk of the stellar populations in them are even more bottom heavy than the Salpeter IMF which is generally believed to be already more bottom heavy than the reality in most environments. This obviously provoked the community greatly resulting in a cascade of many papers on the issue (e.g., Spiniello *et al.* 2011; Worthey *et al.* 2011; Capellari *et al.* 2012; Ferreras *et al.* 2013; Smith 2014).

IMF has been one of those scary contents in the Pandora chest of astronomy along with black holes, dark matter, binaries, and turbulence. Whenever we are incapable of explaining something we are tempted to open the chest and see if any of the contents can help solve the problem. But this is usually the final choice where nothing else can be done, because, although their presence is not doubted, their abundance and effects are not clearly understood at all while their side effects are formidable.

van Dokkum and Conroy used mainly NaI 8200 in their investigation. It is in a tough region of wavelength for measurement because telluric contamination is difficult to remove. NaD 5900 is on the other hand much stronger and easy to measure being in the optical band. A large spectroscopic survey such as SDSS provides its strength for millions of galaxies. So, we have performed a sanity check whether the same models that were used to match the NaI 8200 of bright ellipticals can also reproduce NaD 5900 on the same types of galaxies. Much of the result of our investigation has already been published (Jeong *et al.* 2013) and readers are asked to visit the paper for further details.

2. NaD Excess Objects (NEOs)

We found many bright galaxies do show strong NaD 5900 lines from SDSS. We first constructed a volume-limited sample ($z < 0.08$, $M_r < -20.5$, S/N ¿ 20) of galaxies and secured a better subset of data using the spectral fitting quality assessment of the OSSY database (Oh *et al.* 2011). OSSY provides new and improved measurements of all stellar absorption and emission lines of SDSS galaxies. We found that 1603 galaxies (7.8% of all these bright galaxies) show strong NaD lines and thus were named ?NaD Excess Objects? (NEOs). We were able to morphological classify 963 of them (others were very uncertain) into 553 early-type and 410 late-type galaxies.

3. Late-type NEOs

Late-type NEOs are not an issue here and so are only briefly summarized as follows. Compared to the control sample, most late-type NEOs seem more actively forming stars and dusty. Their NaD strengths are too strong to be of stellar origin, and it is natural to guess that they are of non-stellar origin, for example, due to interstellar matter processes. Similar conclusions were reported by Chen *et al.* (2010).

4. Early-type NEOs

We found it necessary to separate early-type NEOs into (morphologically-)peculiar NEOs and (morphologically-)smooth NEOs because their properties were significantly different from each other.

4.1. *Morphologically-peculiar early-type NEOs*

First on the peculiar NEOs. They often show merger-related features or dust lanes. Compared to their control sample (of ordinary non-NEO elliptical galaxies) they appear to be flatter and dustier, that can be a hint of disky components. They occupy a wide range of stellar mass with no clear preference of mass. As can be guessed from their dusty nature, they show star-formation hinting (by BPT diagnostics) emission lines much more frequently than the control sample does. Their Balmer line based (and thus luminosity-weighted) ages are much younger (roughly 3Gyr) compared to those of the control sample (roughly 10Gyr). Just like late-type NEOs, their NaD strengths go outside the grids of stellar population models. In essence, peculiar early-type NEOs are much like late-type NEOs. And it is not unnatural to suspect a non-stellar origin for their strong NaD lines.

4.2. *Morphologically-smooth early-type NEOs*

These are the ones that are similar to the galaxies studied by van Dokkum and Conroy and hence of interest. Compared to the control sample, they look normal in every sense. There is no peculiarity in morphology. They look round and the internal extinction (adopted from OSSY) appears negligibly low. There is strong mass effect: the more massive the stronger NaD absorption line, in good agreement with van Dokkum and Conroy (2012). Only a small fraction (¡10%) shows emission lines, but even then, most of them indicate LINER origin rather than star formation. Their integrated colours are redder. The absorption line analysis suggests super solar in metals and alpha elements.

Their luminosity-weighted ages are as old as those of the control sample. Our new analysis shows no notable doppler signature on the NaD lines, indicating no inflow-outflow related origins (Park, Jeong and Yi in prep.). In every sense, they are showing stellar lights of typical old populations. Therefore, the strong Na lines should also be of stellar origin.

5. Origin of smooth early-type NEOs

We now focus only on smooth-looking early-type NEOs. Figure 21 of Jeong *et al.* (2013) sums up the situation as to whether the current models can reproduce the observed strengths of NaD lines of smooth early-type NEOs. From our absorption line analysis, it seems almost certain that smooth early-type NEOs are more metal rich in Fe and alpha elements probably by as much as [Fe/H] $\sim +0.3$ and [α/Fe] $\sim +0.3$. Both of these overabundances make NaD lines stronger but not as much as found in smooth early-type NEOs. We need an extra enhancement in Na by as much as [Na/Fe] $\sim +0.3$ to fully account for the strengths of NaD lines observed. The use of bottom heavy IMF hardly helps the situation. It enhances NaD line strength but almost insignificantly, even if we use Conroy?s models. As a result, the rescue plan of bottom heavy IMF that was proposed to solve the mystery of strong NaI 8200 seems futile when it comes to NaD 5900.

6. Discussion

At this conference, Conroy presented a talk and also confirmed publicly that the use of bottom heavy models did not reproduce NaD line strengths. Only with an additional and *ad hoc* Na enhancement of [Na/Fe] $\sim +0.3$ was he able to reproduce the observation, which is consistent with what we have found. We are pleased with this agreement, but that means that the enigma of Na line excess is still an outstanding issue. There is little doubt that Na line strengths contain critical information on IMF and other details of galaxy formation, but whether they are truly indicative of extra bottom-heavy IMF as originally proposed is doubtful.

References

Chen, Y.-M., Tremonti, C. A., Heckman, T., Kauffmann, G., Weiner, B., Brinchmann, J., & Wang, J., 2010, *AJ*, 140, 445

Dutton, A. A., Maccio, A. V., Mendel, J. T., & Simard, L., 2013, *MNRAS*, 432, 2496

Ferreras, I., La Barbera, F., de la Rosa, I. G., Vazdekis, A., de Carvalho, R. R., Falcon-Barroso, J., & Ricciardelli, E., 2013, *MNRAS*, 429, 15

Hardy, E. & Coutour, J., 1988, *ApJ*, 325, 29

Jeong, H., Yi, S. K., Kyeong, J., Sarzi, M., Sung, E.-C., & Oh, K., 2013, *ApJS*, 280, 7

Oh, K., Sarzi, M., Schawinski, K., & Yi, S. K., 2011, *ApJS*, 195, 13

O'Connell, R. W., 1976, *ApJ*, 206, 370

Peterson, R., 1976, *ApJ*, 210, 123

Smith, R., 2014, *MNRAS*, 443, 69

van Dokkum, P. & Conroy, C., 2010, *Nature*, 468, 940

van Dokkum, P. & Conroy, C., 2011, *ApJ*, 735, 13

van Dokkum, P. & Conroy, C., 2012, *ApJ*, 760, 70

Worthey, G., Ingermann, B. A., & Serven, J., 2011, *ApJ*, 729, 148

Galaxy Masses as Constraints of Formation Models
Proceedings IAU Symposium No. 311, 2014
M. Cappellari & S. Courteau, eds.

doi:10.1017/S1743921315003427

Scaling Laws for Dark Matter Halos in Late-Type and Dwarf Spheroidal Galaxies

John Kormendy[1] and K. C. Freeman[2]

[1]Department of Astronomy, University of Texas at Austin,
2515 Speedway, Mail Stop C1400, Austin, Texas 78712-1205, USA,
email: kormendy@astro.as.utexas.edu;
Max-Planck-Institut für Extraterrestrische Physik,
Giessenbachstrasse, D-85748 Garching-bei-München, Germany;
Universitäts-Sternwarte, Scheinerstrasse 1, D-81679 München, Germany;

[2]Research School of Astronomy and Astrophysics,
Mount Stromlo Observatory, The Australian National University,
Cotter Road, Weston Creek, Canberra, ACT 2611, Australia,
email: Kenneth.Freeman@anu.edu.au

Abstract. Dark matter (DM) halos of Sc–Im galaxies satisfy structural scaling laws analogous to the fundamental plane relations for elliptical galaxies. Halos in less luminous galaxies have smaller core radii r_c, higher central densities $\rho_\circ$, and smaller central velocity dispersions σ. If dwarf spheroidal (dSph) and dwarf Magellanic irregular (dIm) galaxies lie on the extrapolations of these correlations, then we can estimate their baryon loss relative to that of Sc – Im galaxies. We find that, if there had been no enhanced baryon loss relative to Sc – Im galaxies, typical dSph and dIm galaxies would be brighter by $\Delta M_B \simeq -4.0$ mag and $\Delta M_B \simeq -3.5$ mag, respectively. Instead, the galaxies lost or retained as gas (in dIm galaxies) baryons that could have formed stars. Also, dSph and dIm galaxies have DM halos that are more massive than we thought, with $\sigma \sim 30$ km s^{-1} or circular-orbit rotation velocities $V_{\rm circ} \sim 42$ km s^{-1}. Comparison of DM and visible matter parameter correlations confirms that, at $M_V \gtrsim -18$, dSph and dIm galaxies form a sequence of decreasing baryon-to-DM mass ratios in smaller dwarfs. We show explicitly that galaxy baryon content goes to (almost) zero at $V_{\rm circ} \lesssim 42 \pm 4$ km s^{-1}, in agreement with $V_{\rm circ}$ as found from our estimate of baryon depletion. Our results suggest that there may be a large population of DM halos that are dark and undiscovered. This helps to solve the problem that the initial fluctuation spectrum of cold dark matter predicts more dwarf galaxies than we observe.

1. Introduction and Analysis Machinery

This paper summarizes Kormendy & Freeman (2014). That paper derives structural parameter correlations for DM halos in Sc–Im and dSph galaxies. We restrict ourselves to objects that contain only two main components, a baryonic disk or main body and a DM halo. For galaxies with well measured H I rotation curves $V(r)$, the derived DM parameters come from published maximum-disk decompositions of $V(r)$ into visible and dark components. The halo model used is the nonsingular isothermal. At absolute magnitude $M_B \gtrsim -14$, V is comparable to the velocity dispersion; then rotation curve decomposition is impossible. For these dwarf spheroidal (dSph) and dwarf Magellanic irregular (dIm) galaxies, we derive central DM densities using the Jeans equation.

We compare DM parameters with visible matter parameter correlations from Kormendy & Bender (2012). They show that the parameter correlations of Sph galaxies are continuous with the disks (but not bulges) of S0 galaxies. In essence, Sph galaxies are bulgeless S0s. Moreover, Sph and S+Im galaxies have similar structure at each M_V. Effective brightnesses decrease dramatically at $M_V > -18$; this is used in Figure 3 here.

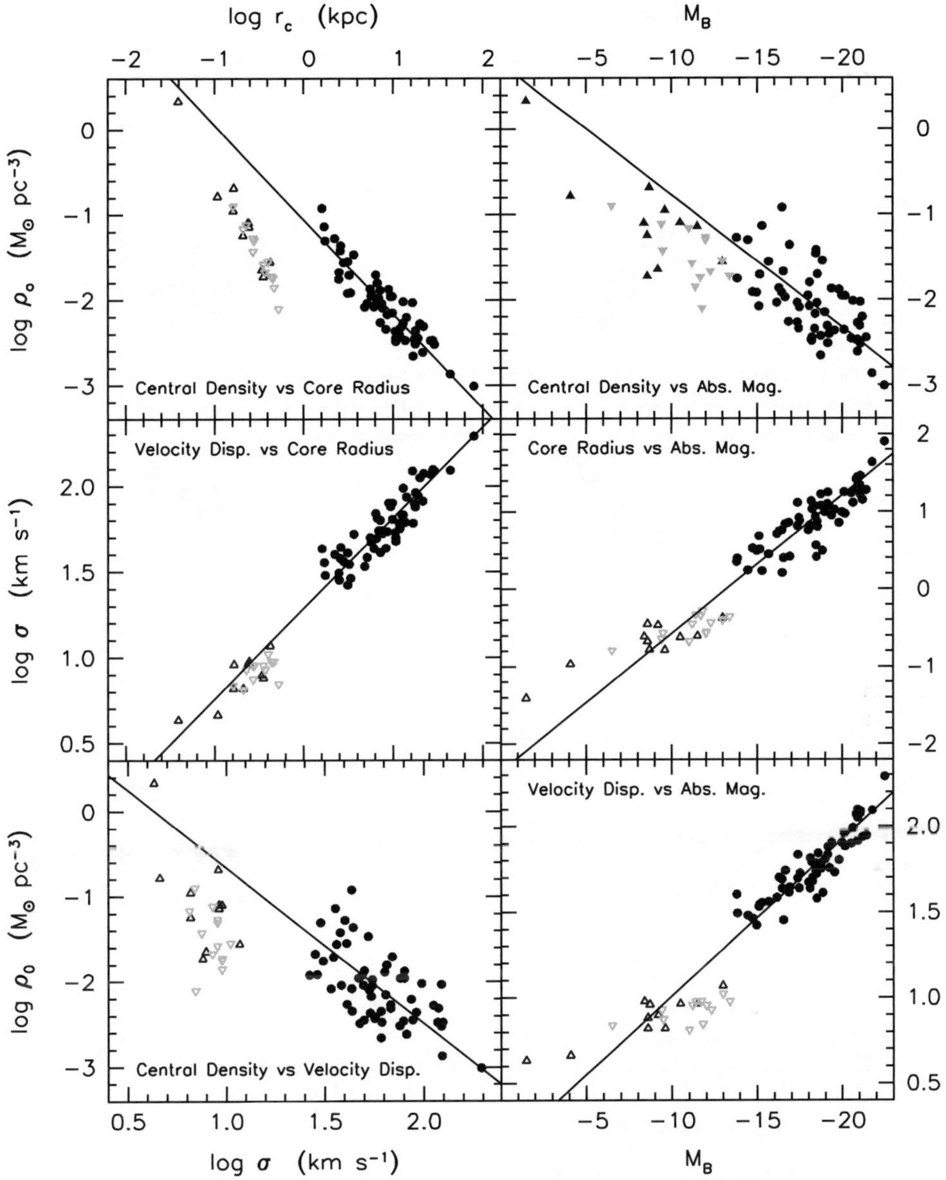

Figure 1. Dark matter parameter correlations for Sc – Im galaxies as derived from rotation curve decompositions using the nonsingular isothermal as DM model (*black points and black line = symmetric least-squares fit*). Also added are dSph galaxies (*green triangles*) and dIm galaxies (*upside-down green triangles*). For dSph and dIm galaxies, $\rho_\circ$ is a meaure of the DM and therefore is plotted with filled symbols in the top-right panel. But r_c and σ are visible-matter parameters and so are plotted with open symbols in the other panels. From Kormendy & Freeman (2014).

2. DM Structural Parameter Correlations

Figures 1 and 2 show the main observational result of Kormendy & Freeman (2014): DM halos of Sc – Im galaxies satisfy well defined scaling laws. Halos in less luminous galaxies have smaller core radii, higher central densities, and smaller velocity dispersions. This confirms previous analyses of smaller samples (Kormendy 1988, 1990; Kormendy & Freeman 2004). Scaling laws provide new constraints on galaxy formation. For example:

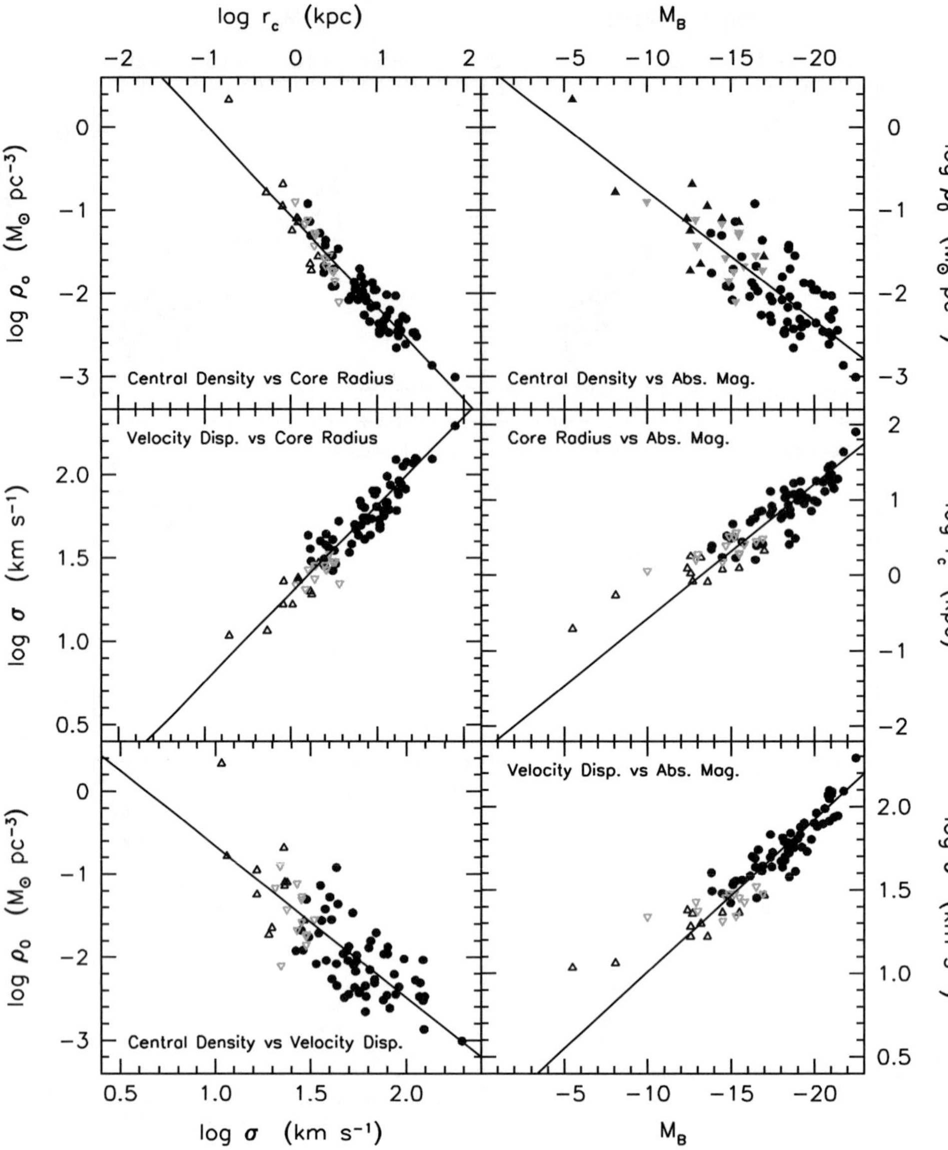

Figure 2. DM correlations for Sc − Im, dSph, and dIm galaxies (Figure 1) but with dSph and dIm galaxies shifted in M_B, $\log r_c$, and $\log \sigma$ (but not $\log \rho_\circ$) to move them onto the scaling laws for the Sc − Im galaxies. The goal is to estimate (1) the baryon loss from dwarf galaxies relative to that from Sc − Im galaxies and (2) the difference between the baryon and DM values of $\log r_c$ and $\log \sigma$. The shifts for dSph and dIm galaxies are, respectively, $M_B \rightarrow M_B - 4.0$ and $M_B \rightarrow M_B - 3.5$, $\log r_c \rightarrow \log r_c + 0.70$ and $\log r_c \rightarrow \log r_c + 0.85$, and $\log \sigma \rightarrow \log \sigma + 0.40$ and $\log \sigma \rightarrow \log \sigma + 0.50$. Note the interpretation of the shifts: $\Delta \log r_c$ gives us the ratio of DM core radius to visible matter core radius, and $\Delta \log \sigma$ gives us the ratio of DM velocity dispersion to visible matter velocity dispersion. From Kormendy & Freeman (2014).

Halo density depends on collapse redshift $z_{\rm coll}$ as $\rho_\circ \propto (1 + z_{\rm coll})^3$. Thus $\rho_\circ$ increases toward lower luminosities because fainter galaxies collapsed earlier. The DM correlations imply that dwarf galaxies formed at least $\Delta\, z_{\rm coll} \simeq 7$ earlier than giant spirals. Correction for baryonic DM compression would make the "pristine" $\rho_\circ$ smaller for giant galaxies and would increase $\Delta\, z_{\rm coll}$.

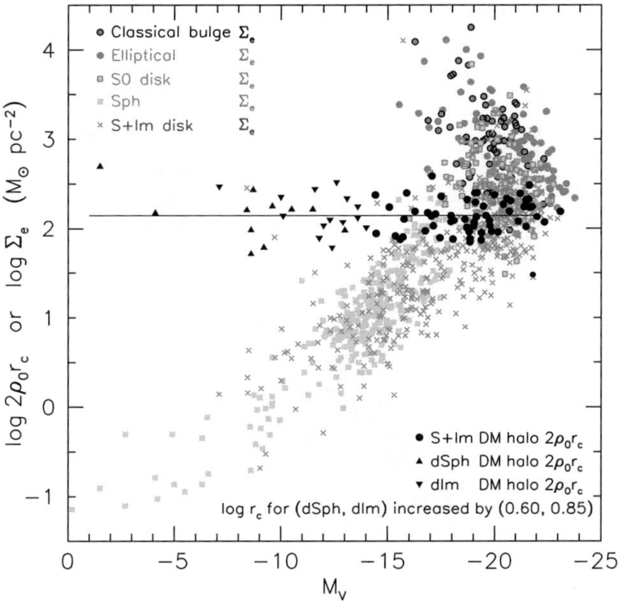

Figure 3. Comparison of DM halo parameters from Fig. 2 with visible matter galaxy parameters from Kormendy & Bender (2012). Central projected densities are plotted for DM halos; effective surface densities $\Sigma_e = \Sigma(r_e)$ are shown for visible components. Here r_e is the radius that encloses half of the light of the component. To convert surface brightnesses to stellar surface densities, we assume mass-to-light ratios $M/L_V = 8$ for ellipticals, 5 for bulges and S0 disks, and 2 for spiral galaxy disks, Im galaxies, and Sph galaxies. From Kormendy & Freeman (2014).

In Figure 3, smaller dwarfs have smaller stellar-to-DM ratios; these are probably associated with baryon loss. The correlations in Figure 1 provide a way to estimate this baryon loss and the properties of dwarf galaxy halos. The Jeans equation tells us the DM central density $\rho_\circ$ but not its core radius r_c or velocity dispersion σ. So the top-right panel in Figure 1 correctly shows an offset of dSph+dIm galaxies from the extrapolation of the fitted $\rho_\circ$–M_B correlation. But the smallest galaxies with rotation curve decompositions have DM densities similar to those of the biggest dwarfs with Jeans equation estimates. *We assume that dSph+dIm galaxies would lie on the extrapolation of the* $\log \rho_\circ - M_B$ *correlation for bright galaxies except for the effects of their enhanced baryon loss.* The top-right panel of Figure 1 then tells us the ΔM_B shifts that bring dSph and dIm galaxies onto the fitted relation. The top-left and bottom-left panels tell us the shifts in $\log r_c$ and $\log \sigma$ that bring dSph+dIm galaxies onto those relations. These shifts are applied in Figure 2. In all panels, dSph and dIm galaxy halos lie on the correlations for more massive galaxies. The $\Delta \log \sigma$ shifts imply that almost-dark dwarfs are more massive than we thought. Their typical halo has $\sigma \sim 30$ km s^{-1}. This corresponds to $V_{\rm circ} \sim 42$ km s^{-1}, in remarkably good agreement with the value of $V_{\rm circ}$ where galaxies get dim in Figure 4.

3. Comparison of Scaling Relations for Visible and Dark Matter

Figure 3 compares the central projected densities of DM halos with effective projected densities of stars. S+Im+Sph galaxies with $M_V \gtrsim -18$ form a sequence of decreasing baryon-to-DM density ratios at decreasing L_V. We suggest that they form a sequence of decreasing baryon retention (vs. supernova-driven winds: Dekel & Silk 1986) or decreasing baryon capture (after cosmic reionization) in smaller galaxies. For galaxies with present

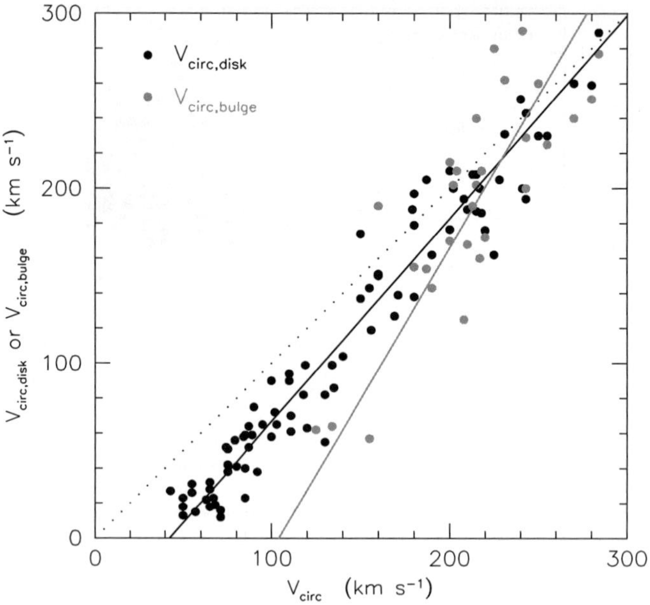

Figure 4. Maximum rotation velocity of the bulge $V_{\rm circ,bulge}$ (*red points*) and disk $V_{\rm circ,disk}$ (*black points*) given in bulge-disk-halo decompositions of galaxy rotation curves whose outer, DM test particle rotation velocities are $V_{\rm circ}$. The dotted line indicates $V_{\rm circ,bulge} = V_{\rm circ,disk} = V_{\rm circ}$. Every red point has a corresponding black point, but many galaxies are bulgeless, and then only a disk was included in the decomposition. This figure shows that the "rotation curve conspiracy", $V_{\rm circ,bulge} \simeq V_{\rm circ,disk} \simeq V_{\rm circ}$ for the halo (Bahcall & Casertano 1985; van Albada & Sancisi 1986; Sancisi & van Albada 1987), happens mostly for galaxies with $V_{\rm circ} \sim 200$ km s^{-1}. The lines are least-squares fits with each variable symmetrized around 200 km s^{-1}. The correlation for bulges is steeper than the one for disks; bulges disappear at $V_{\rm circ} \sim 104 \pm 16$ km s^{-1}. Disks disappear robustly at $V_{\rm circ} = 42 \pm 4$ km s^{-1}. From Kormendy & Freeman (2014).

$M_V \sim -10 \pm 3$, the baryon depletion is estimated by ΔM_B in the caption of Figure 2. If the stellar M/L_V is similar in dwarfs and Sc – Im systems, then ΔM_B measures the mass of stars in dwarfs *relative to the mass of stars in Sc–Im galaxies with halos of similar* $\rho_\circ$. This agrees with their mass-to-light ratios, $M/L_V \sim 10^2$. These dwarfs are almost dark, probably because they lost more baryons than the (also substantial) loss from Sc – Im galaxies. In dIm galaxies, some baryons are still in cold gas, but this effect is fairly small.

Also in Figure 3, projected (not volume!) DM density $\Sigma_{DM\circ}$ is essentially independent of galaxy luminosity L_V. This is a well known result (Kormendy & Freeman 2004; Spano *et al.* 2008; Gentile *et al.* 2009; Donato *et al.* 2009; Plana *et al.* 2010). It implies a Faber-Jackson (1976) relation of the form DM mass $M_{DM} \propto \sigma^4$.

Finally, in Figure 3, bulges and elliptical galaxies have $\Sigma_e > \Sigma_{DM\circ}$, more so at lower L_V. Kormendy *et al.* (2009) and Kormendy & Bender (2012) suggest that they form a sequence of increasing dissipation in the formation of smaller galaxies.

Perhaps the most remarkable result in Kormendy & Freeman (2014) is shown in Fig. 4. It shows that rotation-curve decompositions reveal a robust, linear correlation between the maximum rotation velocity $V_{\rm circ,disk}$ of baryonic disks and the outer circular velocity $V_{\rm circ}$ of test particles in their DM halos. It explicitly shows that $V_{\rm circ,disk} \rightarrow 0$ km s^{-1} at $V_{\rm circ} > 0$ km s^{-1}. In fact, baryons become unimportant at $V_{\rm circ} = 42 \pm 4$ km s^{-1}. This $V_{\rm circ,disk} = 0$ intercept agrees very well with the typical halo σ that we deduced in Section 2. Smaller galaxies are dim or dark. For example, the two extremely faint dSph galaxies in our sample, Segue 1 and Coma, have $V_{\rm circ}$ values of about 16 km s^{-1}.

4. Does There Exist a Large Population of Dark Galaxies?

From the above results, we conclude that the range of visible matter content of $\sigma \sim 30$ km s^{-1} DM halos is large. The ones with the most baryons rotate enough to allow rotation curve decomposition. The ones with the fewest baryons are barely discoverable. None of these galaxies "know" that they must retain $\sim 1\,\%$ of their baryons to be discoverable by us almost 14 billion years after they formed. Moreover, as luminosity decreases toward barely discoverable galaxies, these dwarfs become much more numerous as well as much more nearly dominated by DM. And baryon depletion processes should be more efficient in smaller galaxies. All this suggests that there may exist a large population of objects that are even darker – too dark to be discovered by current techniques. This would help to solve the problem that such objects are predicted by cold DM theory but not seen in Local-Group-like environments (Moore *et al.* 1999; Klypin *et al.* 1999).

Acknowledgments

JK thanks the Directors and staff of Mt. Stromlo Observatory (Australia), of the Observatory of the Ludwig-Maximilians-University (Munich, Germany), and of the Max-Planck-Institute for Extraterrestrial Physics (Garching-by-Munich, Germany) for hospitality and support during many visits over many years when this work was done. We made extensive use of NASA's Astrophysics Data System bibliographic services and of the NASA/IPAC Extragalactic Database (NED). NED is operated by the Jet Propulsion Laboratory and Cal Tech under contract with NASA. And we used the HyperLeda electronic database at `http://leda.univ-lyon1.fr` (Paturel *et al.* 2003). JK's work was supported by NSF grants AST-9219221 and AST-0607490, by the Alexander von Humboldt-Stiftung (Germany), and by Sonderforschungsbereich 375 of the German Science Foundation. The visits to Mt. Stromlo were made possible by the long-term support provided to JK by the Curtis T. Vaughan, Jr. Centennial Chair in Astronomy.

References

Bahcall, J. N., & Casertano, S. 1985, *ApJ*, 293, L7

Dekel, A., & Silk, J. 1986, *ApJ*, 303, 39

Donato, F., Gentile, G., Salucci, P., *et al.* 2009, *MNRAS*, 397, 1169

Faber, S. M., & Jackson, R. E. 1976, *ApJ*, 204, 668

Gentile, G., Famaey, B., Zhao, H., & Salucci, P. 2009, *Nature*, 461, 627

Klypin, A., Kravtsov, A. V., Valenzuela, O., & Prada, F. 1999, *ApJ*, 522, 82

Kormendy, J. 1988, in *Guo Shoujing Summer School of Astrophysics; Origin, Structure and Evolution of Galaxies*, ed. Fang Li Zhi (Singapore: World Scientific), 252

Kormendy, J. 1990, in *The Edwin Hubble Centennial Symposium: Evolution of the Universe of Galaxies*, ed. R. G. Kron (San Francisco: ASP), 33

Kormendy, J., & Bender, R. 2012, *ApJS*, 198, 2

Kormendy, J., Fisher, D. B., Cornell, M. E., & Bender, R. 2009, *ApJS*, 182, 216

Kormendy, J., & Freeman, K. C. 2004, in *IAU Symposium 220*, Dark Matter in Galaxies, ed. S. D. Ryder, D. J. Pisano, M. A. Walker, & K. C. Freeman (San Francisco: ASP), 377

Kormendy, J., & Freeman, K. C. 2014, *ApJ*, submitted (arXiv:1411.2170)

Moore, B., Ghigna, S., Governato, F., *et al.* 1999, *ApJ*, 524, L19

Paturel, G., Petit, C., Prugniel, P., *et al.* 2003, *A&A*, 412, 45

Plana, H., Amram, P., Mendes de Oliveira, C., & Balkowski, C. 2010, *AJ*, 139, 1

Sancisi, R., & van Albada, T. S. 1987, in *IAU Symposium 117, Dark Matter in the Universe*, ed. J. Kormendy & G. R. Knapp (Dordrecht: Reidel), 67

Spano, M., Marcelin, M., Amram, P., *et al.* 2008, *MNRAS*, 383, 297

van Albada, T. S., & Sancisi, R. 1986, *Phil. Trans. R. Soc. London A*, 320, 447

Galaxy Masses as Constraints of Formation Models
Proceedings IAU Symposium No. 311, 2014
M. Cappellari & S. Courteau, eds.

doi:10.1017/S1743921315003439

Angular Momentum across the Hubble sequence from the CALIFA survey

Jesús Falcón-Barroso[1,2], **Mariya Lyubenova**[3,4], **Glenn van de Ven**[4] **and the CALIFA collaboration**

[1]Instituto de Astrofísica de Canarias, Vía Láctea s/n, La Laguna, Tenerife, Spain
[2]Departamento de Astrofísica, Univ. de La Laguna, E-38205 La Laguna, Tenerife, Spain. email: `jfalcon@iac.es`
[3]Kapteyn Astronomical Institute. University of Groningen Postbus 800, 9700 AV Groningen, The Netherlands. email: `lyubenova@astro.rug.nl`
[4]Max Planck Institute for Astronomy, Königstuhl 17, 69117 Heidelberg, Germany. email: `glenn@mpia.de`

Abstract. We investigate the stellar angular momentum of galaxies across the Hubble sequence from the CALIFA survey. The distribution of CALIFA elliptical and lenticular galaxies in the $\lambda_{\rm Re} - \epsilon_{\rm e}$ diagram is consistent with that shown by the Atlas$^{\rm 3D}$ survey. Our data, however, show that the location of spiral galaxies in this diagram is significantly different. We have found two families of spiral galaxies with particularly peculiar properties: (a) spiral galaxies with much higher $\lambda_{\rm Re}$ values than any elliptical and lenticular galaxy; (b) low-mass spiral galaxies with observed $\lambda_{\rm Re}$ values much lower than expected for their apparent flattening. We use these two families of objects to argue that (1) fading alone cannot explain the transformation of spiral to lenticular galaxies, and (2) that those low-mass spiral galaxies are in fact dark matter dominated, which explains the unusually low angular momentum.

Keywords. galaxies: elliptical and lenticular, cD - galaxies: evolution - galaxies: formation

1. The CALIFA survey

The CALIFA survey (Sánchez *et al.* 2012) provides a morphologically unbiased and statistically well defined view of the stellar and ionised gas properties of up to 600 galaxies in the redshift range $0.005 < z < 0.03$. The long wavelength coverage (3400–7300 Å) allows an accurate determination of the stellar population and ionised gas properties (Marino *et al.* 2013, Pérez *et al.* 2013).

In the work we present here we use the dedicated set of high-spectral resolution observations over 3750–4550 Å with the PPAK V1200 grating to derive high-quality stellar kinematics for a sample of 300 galaxies from elliptical E to spiral Sd morphological types. The wide field-of-view of the PMAS/PPAK instrument routinely covers well beyond the effective radius of the observed galaxies.

2. Angular momentum across the Hubble sequence

The combination of the apparent specific stellar angular momentum $\lambda_{\rm Re}$ and the ellipticity $\epsilon_{\rm e}$ (Fig. 1) is an often used diagnostic tool to constrain the dynamical structure and evolution of galaxies. If all galaxies would be simple oblate isotropic rotators they should follow the red curve or lie to the left of it when viewed away from edge-on. The green curve indicates the demarcation line between fast rotator (FR, circles) and slow rotator (SR, squares) galaxies as inferred from the Atlas$^{\rm 3D}$ survey of elliptical (E) and lenticular (S0) galaxies (grey symbols, Emsellem *et al.* 2011).

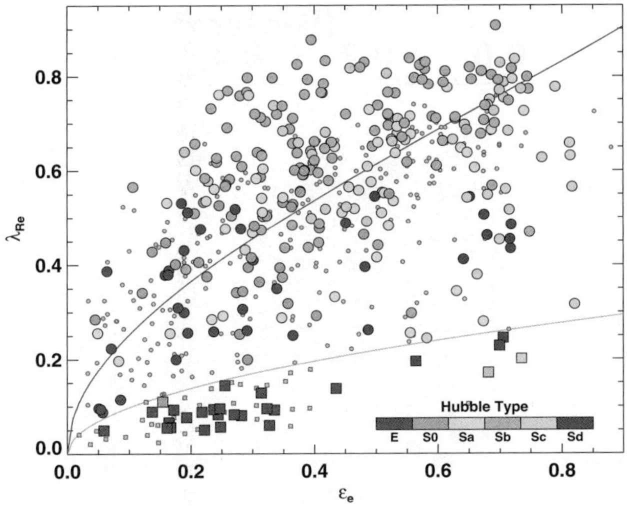

Figure 1. Apparent stellar angular momentum λ_{Re} as function of the ellipticity ϵ_{e} for 300 CALIFA galaxies color-coded by their morphological type. The green curve indicates the demarcation line between slow-rotator (squares) and fast- rotator galaxies (circles) as inferred from the Atlas$^{3\mathrm{D}}$ survey of elliptical (E) and lenticular (S0) galaxies, which are plotted as grey symbols (Emsellem *et al.* 2011). The red curve indicates the location in case galaxies were simple isotropic oblate rotators viewed edge-on.

With the CALIFA survey we are exploring for the first time in a homogeneous way all Hubble types. Spiral galaxies, as expected, are nearly all fast rotator galaxies. However, in comparison with E and S0 galaxies, especially the Sb galaxies reach very high λ_{Re} values, while several Sc and Sd galaxies exhibit low λ_{Re} values and fall just inside the slow rotator regime, in both cases having interesting consequences as discussed in the following sections.

3. Are lenticular galaxies faded spirals?

The angular momentum is set very early on in the life of galaxies, providing an important tool to distinguish between different scenarios of galaxy evolution. For example, a possible explanation of the observed morphology-density relation is the transformation of spiral into fast rotating lenticular galaxies through fading of the stellar populations (e.g. Cappellari *et al.* 2011). Under this scenario the light concentration and the angular momentum are not expected to change significantly.

We measured the apparent specific stellar angular momentum within the half-light radius λ_{Re} (left axis in Fig. 2) for our sample of 300 CALIFA galaxies. The light concentration based on SDSS r-band Petrosian radii (bottom axis) is a proxy for the relative flux in a spheroid (bulge+bar) and disk component (indicated in the top axis, following Gadotti 2009). We observe that Sa galaxies indeed overlap with S0/FR galaxies. Sc and Sd galaxies also have low λ_{Re} values, but their concentrations are too small, which taken together with their large gas supplies, makes it difficult to transform them into red sequence S0/FR galaxies by fading alone. Most of the Sb galaxies are in the green valley and hence their gas may have already been quenched. However, their λ_{Re} values are twice as large and light concentrations significantly smaller than S0/FR galaxies today. This hints to a picture in which the red sequence is being built over time from galaxies that contain disks that are substantially larger and have higher angular momenta.

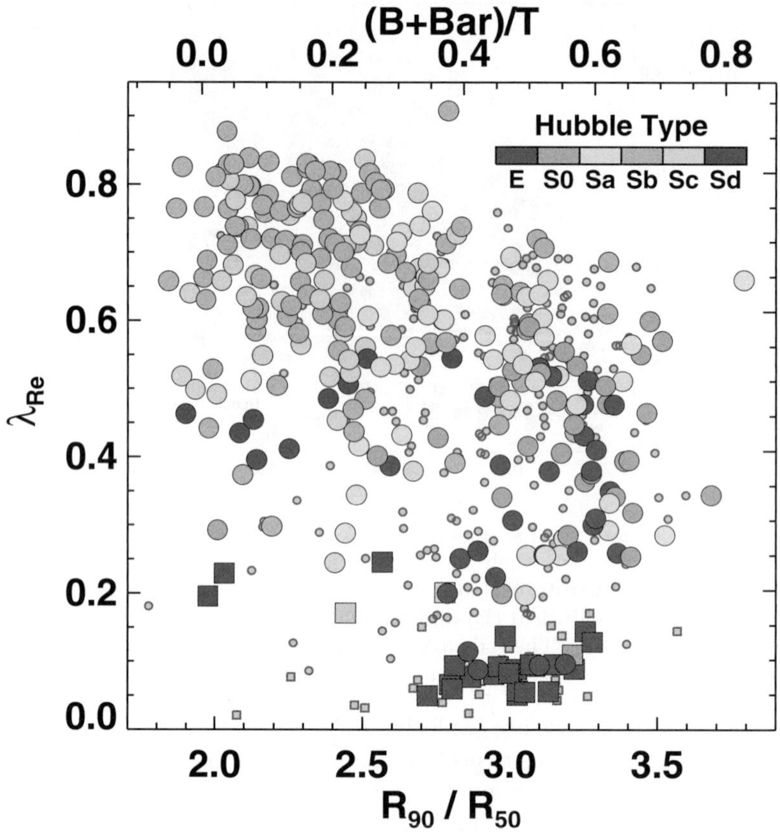

Figure 2. Apparent stellar angular momentum $\lambda_{\rm Re}$ versus light concentration R_{90}/R_{50} based on SDSS r-band Petrosian radii (bottom axis) as a proxy for the relative flux in a spheroid (bulge+bar) and disk component (top axis). Symbols and colors as in Fig.1.

4. Are low-mass spiral galaxies dark matter dominated?

The intrinsic mass distribution of a galaxy is one of the most consequential properties when it comes to its structure and evolution. However, the interplay between baryons and dark matter content is still under heated debate. We inferred the total mass of the 300 CALIFA galaxies of all morphological types by constructing axisymmetric dynamical Jeans models that fit the observed motions of their stars. The stellar mass was derived by fitting the galaxy spectra to stellar population models under the assumption of a Chabrier initial stellar mass function (IMF; González-Delgado *et al.* 2014). The resulting ratio of total-to-stellar mass inside the effective radius (R_e) is shown in Fig. 3 (vertical axis), with arrows indicating the approximate contribution of gas to the baryonic mass (Papastergis *et al.* 2012). On the high-mass end, the upturn above unity could be due to the presence of dark matter or a change toward an IMF even bottom-heavier than the Salpeter IMF as claimed in previous studies (e.g., Cappellari *et al.* 2012, La Barrera *et al.* 2013). However, for low-mass Sc-Sd galaxies there are no indications of a significant deviation from a Chabrier IMF, so that the high total-to-stellar mass ratios are believed to be the result of a high dark matter fraction even within R_e. This is further supported by their observed low $\lambda_{\rm Re}$ values, as the presence of a relatively large dark matter halo would support a dynamically hot but geometrically thin stellar disk.

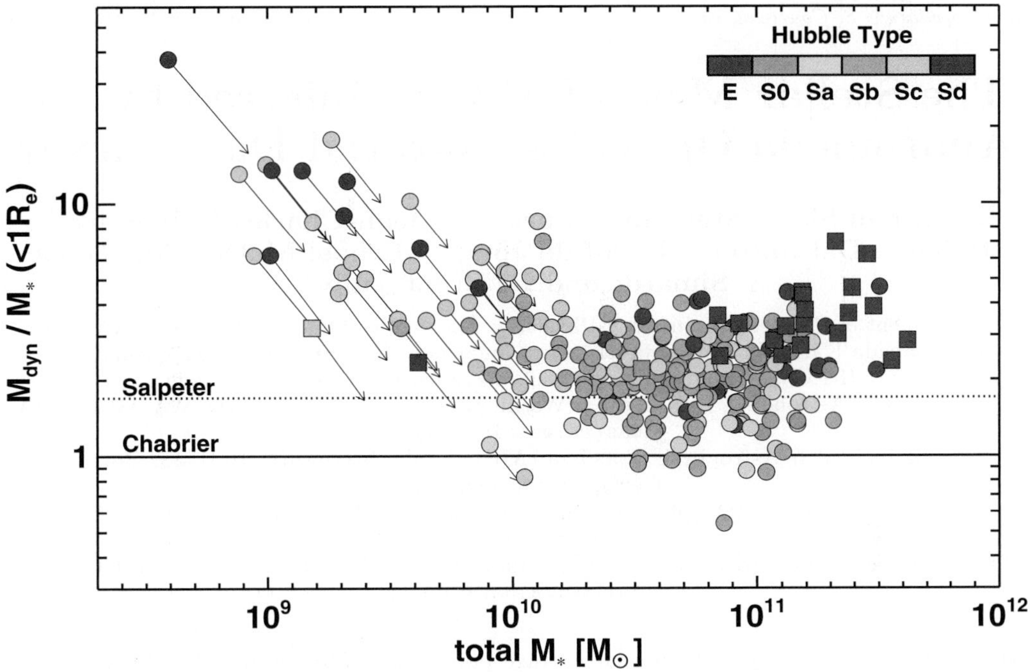

Figure 3. Ratio of total dynamical mass over stellar mass within the half-light radius versus total stellar mass for our sample of CALIFA galaxies. Horizontal lines mark normalizations for different stellar initial mass functions. Symbols and colors as in Fig.1.

Acknowledgements

The authors are grateful to the organizers for the very stimulating meeting celebrating such a special occasion. We are also indebted to the rest of the CALIFA collaboration for the strong support and assistance to carry out this work. JFB acknowledges support from grants AYA2010-21322-C03-02 and AIB-2010-DE-00227 from the Spanish Ministry of Economy and Competitiveness (MINECO). JFB and GvdV acknowledge support from FP7 Marie Curie Actions of the European Commission, via the Initial Training Network DAGAL under REA grant agreement number 289313.

References

Cappellari, M., *et al.*, 2011, *MNRAS*, 413, 813

Cappellari, M., *et al.*, 2012, *Nature*, 484, 485

Emsellem, E., *et al.*, 2011, *MNRAS*, 414, 888

Gadotti, D. A., 2009, *MNRAS*, 393, 1531

González Delgado, R. M., *et al.*, 2014, *A&A*, 562, A47

La Barbera, F., Ferreras, I., Vazdekis, A., de la Rosa, I. G., de Carvalho, R. R., Trevisan, M., Falcón-Barroso, J., & Ricciardelli, E., 2013, *MNRAS*, 433, 3017

Marino, R. A., *et al.*, 2013, *A&A*, 559, A114

Papastergis, E., Cattaneo, A., Huang, S., Giovanelli, R., & Haynes, M. P., 2012, *ApJ*, 759, 138

Pérez, E., *et al.*, 2013, *ApJ*, 764, L1

Sánchez, S. F., *et al.*, 2012, *A&A*, 538, A8

Walcher, C. J., *et al.*, 2014, *A&A*, 569, A1

Galaxy Masses as Constraints of Formation Models
Proceedings IAU Symposium No. 311, 2014
M. Cappellari & S. Courteau, eds.

doi:10.1017/S1743921315003440

The Stellar Mass of M31 as inferred by the Andromeda Optical & Infrared Disk Survey

Jonathan Sick,[1] Stephane Courteau,[1] Jean-Charles Cuillandre,[2] Julianne Dalcanton,[3] Roelof de Jong,[4] Michael McDonald,[5] Dana Simard[1] and R. Brent Tully[6]

[1]Department of Physics, Engineering Physics & Astronomy, Queen's University, Kingston, ON, Canada K7L 3N6. email: jsick@astro.queensu.ca, courteau@astro.queensu.ca
[2]IRFU, Centre d'études de Saclay, France jcc@cfht.hawaii.edu
[3]Department of Astronomy, University of Washington, Box 351580, Seattle, WA 98195, USA. jd@astro.washingston.edu
[4]Leibniz Institut fr Astrophysik Potsdam (AIP), An der Sternwarte 16, 14482 Potsdam, Germany. rdejong@aip.de
[6]Kavli Institute for Astrophysics and Space Research, MIT, Cambridge, MA, USA. mcdonald@space.mit.edu
[6]Institute for Astronomy, University of Hawaii, 2680 Woodlawn Drive, Honolulu, HI, USA. tully@ifa.hawaii.edu

Abstract. Our proximity and external vantage point make M31 an ideal testbed for understanding the structure of spiral galaxies. The Andromeda Optical and Infrared Disk Survey (ANDROIDS) has mapped M31's bulge and disk out to R=40 kpc in $ugriJK_s$ bands with CFHT using a careful sky calibration. We use Bayesian modelling of the optical-infrared spectral energy distribution (SED) to estimate profiles of M31's stellar populations and mass along the major axis. This analysis provides evidence for inside-out disk formation and a declining metallicity gradient. M31's i-band mass-to-light ratio (M/L_i^*) decreases from 0.5 dex in the bulge to $\sim$ 0.2 dex at 40 kpc. The best-constrained stellar population models use the full $ugriJK_s$ SED but are also consistent with optical-only fits. Therefore, while NIR data can be successfully modelled with modern stellar population synthesis, NIR data do not provide additional constraints in this application. Fits to the gi-SED alone yield M/L_i^* that are systematically lower than the full SED fit by 0.1 dex. This is still smaller than the 0.3 dex scatter amongst different relations for M/L_i via $g - i$ colour found in the literature. We advocate a stellar mass of $M_*(30\text{ kpc}) = 10.3^{+2.3}_{-1.7} \times 10^{10}$ M$_\odot$ for the M31 bulge and disk.

Keywords. galaxies: spiral - galaxies: stellar content - galaxies: photometry

1. Introduction

The ANDROIDS programme has used the MegaCam and WIRCam cameras on the Canada-France-Hawaii Telescope (CFHT) to map M31's bulge and disk homogeneously within $R = 40$ kpc with $ugriJK_s$ bands and enable global studies of M31's structure and stellar populations using both resolved stars and integrated spectral energy distributions (SEDs). In this contribution, we use ANDROIDS to estimate the stellar mass profile of the M31 disk with Bayesian modelling of the optical to near-IR (NIR) SED. This approach is more rigorous than the colour-M/L^* prescriptions (e.g. Zibetti *et al.* 2009; Taylor *et al.* 2011; Into & Portinari 2013) often employed by pixel-by-pixel stellar mass estimation studies that use only a $g - i$ colour and marginalize over all likely star formation histories. By studying M31 in detail, an overall goal of ANDROIDS is to explore systematic uncertainties in studies of more distant and poorly resolved systems.

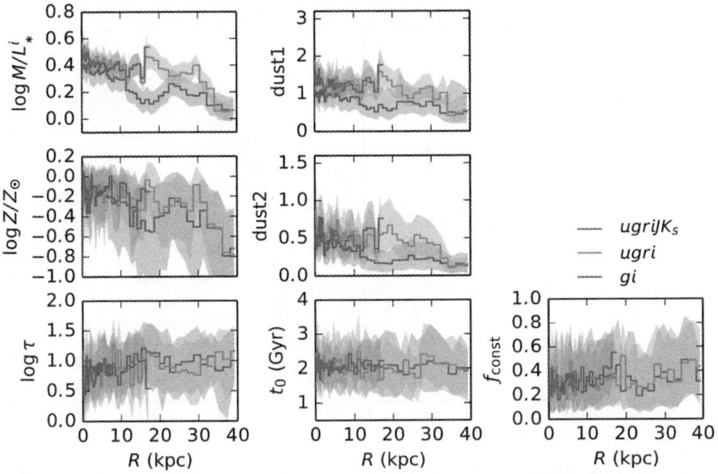

Figure 1. Posterior stellar population profiles for different bandpass sets: $ugriJK_s$ (blue), $ugri$ (green), gi (red). A declining metallicity gradient and inside-out disk formation (seen by an increase in the e-folding time, $\log \tau$, of the exponentially declining star formation history model) are clearly evident.

2. M31 Surface Brightness Calibration

Background subtraction is the most significant challenge for observational studies of M31's structure since we cannot observe its disk and blank sky simultaneously. This is particularly acute in our NIR maps where skyglow is 3-dex brighter than the disk, while also having strong spatial and temporal variations. In Sick *et al.* (2014), we describe our ANDROIDS/WIRCam sky-target nodding and background subtraction schemes and find that the NIR background cannot be known to better than 2% given the scale of sky-target nods required for M31. We overcome this uncertainty by solving for sky offsets that formally minimize surface brightness differences between overlapping pairs of images. Such sky offsets are $\sim$ 1% of the NIR brightness, but systematically uncertain up to a zeropoint normalization that is 0.16% of the sky level. In optical bands, the sky background is both more stable and somewhat dimmer, though we still employ sky-target nodding with the Elixir-LSB method for CFHT/MegaCam to build a real-time map of sky and scattered light backgrounds over one-hour sliding windows. With Elixir-LSB we easily identify low surface brightness features in M31's outer disk, such as the Northern Spur, at levels below $\mu_g \sim 26$ mag arcsec^{-2} (Sick *et al.* 2013).

The aforementioned sky offset zeropoint uncertainty requires that our surface brightness profiles be finely calibrated against external datasets. Resolved stellar catalogs transformed into surface brightness maps, such as our own WIRCam star catalog, and even Panchromatic Hubble Andromeda Treasury (PHAT; Dalcanton *et al.* 2012), provide a useful dataset up to the limit of completeness corrections. Extremely wide-field imaging is also useful as it enables a simultaneous mapping of the background and the disk light. We are currently using Dragonfly (Abraham & van Dokkum 2014) to image M31 as a replacement for the venerable wide-field plates of Walterbos & Kennicutt (1987).

3. SED Stellar Mass Modelling

From the calibrated surface brightness profiles, we model the SED at each radial bin to estimate the stellar population, and hence the stellar mass-to-light ratio, M/L_i^*. Our modelling engine is the Flexible Stellar Population Synthesis (FSPS) software (Conroy

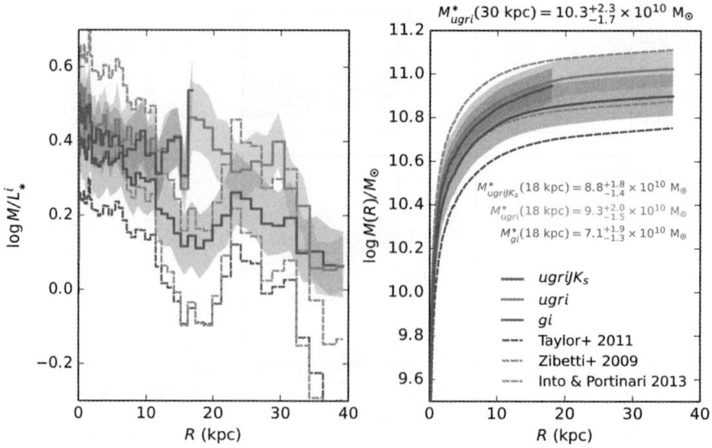

Figure 2. Posterior stellar M/L_i^* (left) and stellar mass (right) major axis profiles. The $ugriJK_s$ (blue) and $ugri$ (green) fits are consistent, while fit of only gi (red) are lower by 0.1 dex in M/L_i^*. Equivalent gi–M/L_i^* relations in the literature can vary by 0.3 dex of M/L_i^*.

et al. 2009, 2010). We chose FSPS for its reliable calibration and "lighter" AGB contribution than older SP models (e.g., Bruzual & Charlot 2003), and allowance for deep customization of the computed stellar populations.† We use a Markov Chain Monte Carlo approach to modelling SEDs extracted along the northern major axis of the M31 disk implemented with the `emcee` python package (Foreman-Mackey *et al.* 2013). We tested different star formation history parameterizations and found that a simple 'τ' model, involving constant plus exponentially declining star formation rate components minimized residuals compared to more sophisticated 'delayed τ' and late burst models. Of the dust attenuation treatments, the default power-law attenuation law with separate components for young and older stellar populations also minimized residuals compared to Milky Way or starburst attenuation models.

We found that posterior SED residuals are minimized by fitting the entire $ugriJK_s$ SED. This contradicts Taylor *et al.* (2011) and Zibetti *et al.* (2009) who advocated against using NIR bands in mass estimation due to uncertain AGB treatments of the previous generation of stellar population synthesis models (e.g. Bruzual & Charlot 2003; Maraston 2005). Much like the NIR, the $griJK_s$-SED fit has little predictive power over the crucial u-band. This result should therefore encourage SED modellers to incorporate as many bandpasses as possible, including UV and IR, to obtain the best constraints on stellar populations and masses.

We modelled SEDs extracted from a logarithmically-sized wedge (e.g. Courteau *et al.* 2011, their Fig. 2) to produce stellar population profiles (shown in Fig. 1). Interestingly, the $ugri$-fit and $ugriJK_s$-fit SEDs produce statistically identical stellar population profiles, with the only exception being a slightly tighter posterior credible region from the full-SED fits. Although the consistency of optical and optical-NIR SED fits is reassuring from the perspective of NIR calibrations, it is also disappointing that the NIR data has not produced a remarkably improved posterior stellar population estimate.

Clearly evident is that poorly sampled SEDs can bias results. Fitting only the gi SED (that is, using an input information equivalent to those using colour-M/L^* look-up-tables) clearly biases the posterior stellar population distribution, with significantly

† The lead author (J.S.) contributes to the maintenance of a Python-language wrapper for FSPS: `http://dan.iel.fm/python-fsps`

lower dust opacities and lower mass-to-light ratios. By comparison, we have also plotted mass-to-light ratios predicted by three colour-M/L^* relations (Zibetti *et al.* 2009; Taylor *et al.* 2011; Into & Portinari 2013). These fits systematically vary by 0.3 dex, far larger than the 0.1 dex of internal systematic uncertainty typically claimed by $g-i-M/L^*$ fits Courteau *et al.* (2014). Compared to our full SED fits, modelling of the *gi* SED is less biased than these other M/L^* fits, which are based on other stellar population synthesis models. This serves as reminder that stellar mass estimates remain dominated by prior assumptions such as choices of IMF, dust, and details of AGB treatments, among other concerns.

4. Discussion

We have used $ugriJK_s$ SEDs to map the stellar mass of M31's disk and find a stellar mass, within 30 kpc, of $M^*_{ugri} = 10.3^{+2.3}_{-1.7} \times 10^{10}$ M$_\odot$. This result is consistent with the stellar bulge and disk masses quoted by Tamm *et al.* (2012) (10.1×10^{10} M$_\odot$). Future work will extend this analysis to a full 2D mapping of the M31 stellar mass distribution.

We are matching these stellar mass maps with dynamical tracers of gas and stars to construct a mass model of M31's stellar, gas, and dark matter components (Simard *et al.*, in progress). The DiskFit code (Spekkens & Sellwood 2007) allows us to correct the HI velocity fields of Saglia *et al.* (2010) and Chemin *et al.* (2009) for non-circular motions in M31's central parts. The success of this mass model will be determined by the stability of a dynamical N-body realization.

Acknowledgements

J.S. and S.C. acknowledge support through respective Graduate Scholarship and Discovery grants from the Natural Sciences and Engineering Research Council of Canada. We thank the Canadian Advanced Network for Astronomical Research (CANFAR) for enabling the computing facilities needed for this work.

References

Abraham, R. G. & van Dokkum, P. G. 2014, *PASP*, 126, 55
Bruzual, G. & Charlot, S. 2003, *MNRAS*, 344, 1000
Chemin, L., Carignan, C., & Foster, T. 2009, *ApJ*, 705, 1395
Conroy, C., Gunn, J. E., & White, M. 2009, *ApJ*, 699, 486
Conroy, C., White, M., & Gunn, J. E. 2010, *ApJ*, 708, 58
Courteau, S., Widrow, L. M., McDonald, M., *et al.* 2011, *ApJ*, 739, 20
Courteau, S., Cappellari, M., de Jong, R. S., *et al.* 2014, Rev. Mod. Phys., 86, 47
Dalcanton, J. J., Williams, B. F., Lang, D., *et al.* 2012, *ApJS*, 200, 18
Foreman-Mackey, D., Hogg, D. W., Lang, D., & Goodman, J. 2013, *PASP*, 125, 306
Into, T. & Portinari, L. 2013, *MNRAS*, 430, 2715
Maraston, C. 2005, *MNRAS*, 362, 799
Saglia, R. P., Fabricius, M., Bender, R., *et al.* 2010, *A&A*, 509, A61
Sick, J., Courteau, S., & Cuillandre, J.-C. 2013, 1310.4832
Sick, J., Courteau, S., Cuillandre, J.-C., *et al.* 2014, *AJ*, 147, 109
Spekkens, K. & Sellwood, J. A. 2007, *ApJ*, 664, 204
Tamm, A., Tempel, E., Tenjes, P., Tihhonova, O., & Tuvikene, T. 2012, *A&A*, 546, A4
Taylor, E. N., Hopkins, A. M., Baldry, I. K., *et al.* 2011, *MNRAS*, 418, 1587
Walterbos, R. A. M. & Kennicutt, Jr., R. C. 1987, A&AS, 69, 311
Zibetti, S., Charlot, S., & Rix, H. 2009, *MNRAS*, 400, 1181

Galaxy Masses as Constraints of Formation Models
Proceedings IAU Symposium No. 311, 2014
M. Cappellari & S. Courteau, eds.

doi:10.1017/S1743921315003452

Galaxy Halo Masses from Weak Gravitational Lensing

Rachel Mandelbaum

McWilliams Center for Cosmology, Department of Physics, Carnegie Mellon University, Pittsburgh, PA 15213, USA. email: rmandelb@andrew.cmu.edu

Abstract. In this review, I discuss the use of galaxy-galaxy weak lensing measurements to study the masses of dark matter halos in which galaxies reside. After summarizing how weak gravitational lensing measurements can be interpreted in terms of halo mass, I review measurements that were used to derive the relationship between optical galaxy mass tracers, such as stellar mass or luminosity, and dark matter halo mass. Measurements of galaxy-galaxy lensing from the past decade have led to increasingly tight constraints on the connection between dark matter halo mass and optical mass tracers, including both the mean relationships between these quantities and the intrinsic scatter between them. I also review some of the factors that can complicate analysis, such as the choice of modeling procedure, and choices made when dividing up samples of lens galaxies.

1. Introduction

The field of galaxy formation and evolution seeks to explain the evolutionary history of galaxies, but is handicapped by the difficulties in observing the dark matter field in which the galaxies form and evolve. The connection between galaxies and their host dark matter halos is an essential ingredient in the physics of galaxy formation. One very useful probe of the galaxy-dark matter connection is weak lensing around galaxies, or galaxy-galaxy lensing (e.g., Hoekstra & Jain 2008).

Gravitational lensing, the deflection of light by mass, induces tangential shear distortions in the shapes of background galaxies around foreground galaxies, allowing direct measurement of the galaxy-matter correlation function. This approach has the advantage of being sensitive to *all* matter, independent of its dynamical state. The individual distortions are small (typically of order 0.1%), but by averaging over all foreground ("lens") galaxies within a given subsample selected based on their properties (a process known as "stacking"), and over all the source galaxies behind them, it is possible to obtain a high signal-to-noise measurement of the shear as a function of angular separation from the galaxy, known as the galaxy-galaxy lensing signal.

In the past $\sim$ 15 years, the quantity of imaging data with high-quality galaxy shape information and with some redshift information (either spectroscopic or photometric redshifts) has drastically increased, in large part due to the efforts of major surveys such as the Red-Sequence Cluster Survey (RCS, Gladders & Yee 2005) and RCS2 (Gilbank *et al.* 2011), the Sloan Digital Sky Survey (SDSS, York *et al.* 2000), the Canada-France-Hawaii Telescope Lensing Survey (CFHTLenS, Heymans *et al.* 2012; Erben *et al.* 2013), and the *Hubble Space Telescope* (*HST*) COSMOS Survey (Koekemoer *et al.* 2007; Scoville, *et al.* 2007a; Scoville *et al.* 2007b). The increase in data quantity and quality, plus the appeal of a measurement method that is directly sensitive to the dark matter, has led to tremendous progress in the use of galaxy-galaxy lensing to study the connection between galaxies and matter.

The purpose of this review is to describe some of the studies that have analyzed galaxy-galaxy weak lensing measurements to learn about the masses of dark matter halos around galaxies, and to highlight important lessons learned and unsolved aspects of this problem that require more work in future, for example for the next generation of large lensing surveys that will measure weak gravitational lensing more precisely than has been done with current datasets. The topic of this review is limited to relatively general studies of typical galaxies, leaving aside studies of particular galaxy types using weak gravitational lensing (e.g., Mandelbaum *et al.* 2009; Dutton *et al.* 2010; Reyes *et al.* 2012), studies of the radial profiles of dark matter halos with weak lensing (e.g., Gavazzi *et al.* 2007; Mandelbaum *et al.* 2008; Schulz *et al.* 2010), and studies of group or galaxy cluster dark matter halos (e.g., Hoekstra *et al.* 2013).

2. Theory

Weak gravitational lensing, the deflection of light rays by the mass in matter along the line-of-sight, results in coherent shape distortions (shears) in the background galaxies. The strength of the lensing shear depends on the mass in the lens object (galaxy or cluster), the separation on the sky between the lens and the source object, and the line-of-sight distances to lens and source.

A galaxy-galaxy weak lensing measurement probes the connection between galaxies and matter via the cross-correlation functions $\xi_{\rm gm}(\vec{r})$, which can be related to the projected surface density of matter around the lens galaxies

$$\Sigma(R) = \overline{\rho} \int \left[1 + \xi_{\rm gm}\left(\sqrt{R^2 + \chi^2}\right)\right] d\chi. \tag{2.1}$$

where R is the transverse separation and χ the radial direction over which we are projecting†. The surface density is then related to the observable quantity for lensing, the differential surface density,

$$\Delta\Sigma(R) = \gamma_t(R)\Sigma_c = \overline{\Sigma}(< R) - \Sigma(R), \tag{2.2}$$

in the weak lensing limit, for a matter distribution that is axisymmetric along the line of sight (which is naturally achieved when stacking thousands of lens galaxies to determine their average lensing signal). Typical galaxy-galaxy lensing measurements require substantial numbers of lens and source galaxies in order to average out the random component of the source galaxy shapes, which is typically the dominant source of noise in the measurement. This observable quantity can be expressed as the product of two factors, a tangential shear γ_t and a geometric factor

$$\Sigma_c = \frac{c^2}{4\pi G} \frac{D_S}{D_L D_{LS}} \tag{2.3}$$

where D_L and D_S are angular diameter distances to the lens and source, and D_{LS} is the angular diameter distance between the lens and source.

For isolated galaxies at the center of dark matter halos, the surface density of surrounding matter can be written in terms of the 3d density profile $\rho(r)$ on small scales, as

$$\Sigma(R) = \int_{-\infty}^{\infty} \rho(r = \sqrt{\chi^2 + R^2}) d\chi. \tag{2.4}$$

However, there are additional contributions due to those galaxies that are satellites, for

† We are ignoring the effects from the radial window, which is broad enough that it is not relevant at galaxy scales; see section 2.3 of Guzik & Seljak (2001) for details.

which the host dark matter halo leads to a contribution for R in the hundreds of kiloparsec to 1 Mpc range. On even larger scales, there is a small but significant contribution due to the matter in structures that are physically associated with but not inside of the halo in which the lens galaxy resides (the "2-halo term"). As a result, the interpretation of the lensing signal can be complicated, depending on the local environment of the galaxies (mixture of isolated and central galaxies vs. satellites) and the separations R used (larger scales requires modeling of the term due to large-scale structure). Moreover, the stacking process means that the average profile can be affected not just by mean relationships between mass and observables, but also by scatter in those relationships. Different studies have taken different approaches to these interpretation challenges, as I will describe below.

3. Results

Among the first galaxy-galaxy lensing results with detailed modeling of the mass distributions of large samples of foreground lens galaxies were Hoekstra *et al.* (2005), Mandelbaum *et al.* (2006), and Heymans *et al.* (2006).

Hoekstra *et al.* (2005) used data from the RCS to identify isolated galaxies and measure their lensing signals as a function of stellar mass or luminosity. Due to the use of isolated galaxies, their approach was to fit the signals to an NFW (Navarro *et al.* 1996) profile, ignoring the possibility of contributions from host halos of those galaxies that may be satellites in larger halos, and restricting to small enough scales that the 2-halo term is negligible. They find a relationship between halo mass and luminosity that goes like $M_{\rm halo} \propto L^{1.5}$. This result is based on the best-fit mass for samples with some average luminosity, ignoring scatter between mass and luminosity. They also made these measurements for early and late-type galaxy samples split based on color, and found $M_{\rm halo}/M_*$ higher by a factor of ~ 2 for early types. Dividing $M_*/M_{\rm halo}$ by the cosmological baryon fraction to estimate an efficiency of conversion of baryons to stars,

$$\eta = \frac{M_*}{M_{\rm halo}} \frac{\Omega_{\rm m}}{\Omega_{\rm b}} \tag{3.1}$$

they found $\eta \sim 33\%$ and $\sim 14\%$, respectively, for early and late type galaxies.

Using data from the SDSS Main spectroscopic galaxy sample, Mandelbaum *et al.* (2006) analyzed the host halo mass for early and late-type central galaxies as a function of their stellar mass and luminosity, for galaxies at a typical redshift of $\langle z \rangle \sim 0.1$ (lower than the RCS study). In this work, the split into early versus late types was achieved using a morphological estimator (not color). This work also estimated the satellite fractions purely based on the lensing signal alone. In order to interpret the lensing signals in terms of central and satellite galaxies in a statistical sense, this work used a halo model (e.g., Seljak 2000; Cooray & Sheth 2002) formalism that was tested for these purposes on mock galaxy samples derived from N-body simulations by Mandelbaum *et al.* (2005). The simple halo model that was used had only two free parameters, with the rest fixed to some values selected based on the N-body simulation analysis, and correction factors for scatter in the mass-observable relation were applied to the best-fit masses. The findings for $M_*/M_{\rm halo}$ in this work were consistent with those from RCS for stellar masses above $10^{10} M_\odot$. Below that stellar mass, early and late type galaxies were found to have statistically consistent conversion efficiencies (Eq. 3.1), suggesting that stellar mass is a good tracer of halo mass below $M_* \sim 10^{10} M_\odot$. For a galaxy sample around L_*, the halo model analysis of the lensing results suggests $M_{\rm halo}/L = 79^{+27}_{-24}$ and $41^{+16}_{-17} M_\odot/L_\odot$ (2σ) for early and late types, respectively, with higher values at higher luminosity; results at lower luminosity are too noisy to draw conclusions about trends in that direction.

Using data from the *HST* GEMS survey, Heymans *et al.* (2006) analyzed the lensing signal from high stellar mass galaxies over a long redshift baseline, $0.2 < z < 0.8$. Given the stellar mass limit $\log(M_*/M_\odot) > 10.5$, the majority of the galaxies exhibit early type morphology; however, no explicit morphological split was used. After modeling the signals using a pure NFW profile without modeling of signal due to satellites or halo vs. stellar mass scatter, they find an average conversion efficiency of $\eta = 0.10 \pm 0.03$ (1σ) for the entire sample. These results are consistent with those from lower redshift (Hoekstra *et al.* 2005; Mandelbaum *et al.* 2006) when comparing with samples that have a similar stellar mass range. When splitting into redshift bins within the GEMS sample, they find no statistically significant evolution of the $M_{\rm halo}/M_*$ ratio, though their best-fit relation includes slight evolution in the direction of higher ratio at higher redshift, giving an upper limit in growth of this ratio of $\lesssim 2.5$ from $z = 0.8$ to the present time.

More recently, Leauthaud *et al.* (2012a) carried out a joint analysis of the galaxy-galaxy lensing, galaxy clustering, and abundance of galaxies as a function of stellar mass (without splitting by galaxy type) in the COSMOS survey. While this survey covers a very small area of the sky, its depth provides a long redshift baseline over which to carry out this analysis. This joint analysis used a quite complex formulation of the halo model with many parameters including ones regulating the scatter between the halo and stellar mass, as described in Leauthaud *et al.* (2011). For this particular dataset, the stellar mass function has the highest S/N and therefore dominates the constraints on the halo vs. stellar mass relation, which is provided as a functional form with four parameters determined in three redshift bins. The redshift evolution of this halo vs. stellar mass relation is indicative of downsizing, with the stellar mass at which the star baryon conversion efficiency is maximized decreasing at lower redshifts. One new aspect to this halo model analysis compared to previous ones used for lensing analyses is that it provides constraints on the logarithmic scatter in the stellar mass at fixed halo mass, which varied in the range $\sigma_{\log M_*} = 0.2$–0.25 for the three redshift bins.

The halo model formalism used by Leauthaud *et al.* (2012a) relied on the use of all galaxies at fixed stellar mass to define the abundance. An updated version of the formalism that permits a split into star-forming and passive galaxies was presented and used by Tinker *et al.* (2013), also on the COSMOS data. The results of that work suggested that for massive galaxies ($M_* \gtrsim 10^{10.6} M_\odot$), star-forming galaxies form stars at a rate that roughly matches the growth of their dark matter halos via accretion from $z = 1$ to 0, whereas for quiescent galaxies, the growth of their dark matter halos outpaces star formation, reducing their apparent baryon conversion efficiency η. In contrast, for lower mass galaxies, the halo vs. stellar mass relations are similar for passive and star-forming galaxies at $z < 1$. These findings are qualitatively consistent with the trends from the SDSS data using lensing alone in Mandelbaum *et al.* (2006), but the S/N in COSMOS is superior for lower mass galaxies. Moreover, COSMOS enables tests of redshift evolution, and find a cross-over in the red-blue stellar vs. halo mass relation as a function of redshift, which has not been seen in other surveys (either due to insufficient redshift baseline, or different modeling strategies).

van Uitert *et al.* (2011) used imaging data from the RCS2 together with SDSS spectroscopic redshifts in order to perform a weak lensing study of the halo vs. stellar mass relation, with a similar halo model formalism as that used in Mandelbaum *et al.* (2006). They find a relatively steep scaling of halo mass with luminosity, 2.2 ± 0.1 and 1.8 ± 0.1 for red and blue galaxies, respectively†, which is inconsistent with that found by Hoekstra

† These numbers differ from those in van Uitert *et al.* (2011), and come from a correction that was reported in Velander *et al.* (2014), which used the same halo modeling software.

et al. (2005) in the RCS; however, the different methods of modeling the lensing signals and different selection criteria used could be responsible for this difference. The results are consistent with those from the SDSS analysis by Mandelbaum *et al.* (2006) within the errors. When splitting the sample by stellar mass, van Uitert *et al.* (2011) confirm the findings from the SDSS: below $10^{11} M_{\odot}$, the stellar mass traces halo mass, but above that stellar mass, the halo masses differ for early and late type galaxies.

van den Bosch *et al.* (2013) presented a 9-parameter conditional luminosity function (CLF) formalism including various input from N-body simulations (halo mass function, concentration-mass relation, and others). Cacciato *et al.* (2014) showed that when fitting the parameters of this model to describe low-redshift SDSS samples, the resulting CLF was able to describe lensing signals for higher-luminosity and redshift SDSS and RCS2 samples without further adjustment. This result is non-trivially interesting and suggests that the 9-parameter model may be capturing the key features of the galaxy-dark matter connection. One interesting aspect of this model is that it includes constant scatter in $\log L$ at fixed $\log M_{\rm halo}$, which implies that the scatter in $\log M_{\rm halo}$ at fixed $\log L$ is not constant (given that the slope of the luminosity-mass relation varies). This aspect of the model is similar to the results of the COSMOS analysis from Leauthaud *et al.* (2012a), which described the results in terms of a fixed scatter in $\log M_*$ at fixed $M_{\rm halo}$.

Two works (Hudson *et al.* 2013; Velander *et al.* 2014) have explored the galaxy-galaxy weak lensing signals as a function of lens galaxy properties using CFHTLenS data. The CFHTLenS data cover sufficient area with enough depth that, compared to the SDSS, RCS, and RCS2, the statistical errors in the lensing signal at lower stellar mass and luminosity have decreased significantly†. When fitting the halo mass to a power law in luminosity (stellar mass), Velander *et al.* (2014) find a power law index of 1.32 ± 0.06 and $1.09^{+0.20}_{-0.13}$ ($1.36^{+0.06}_{-0.07}$ and $0.98^{+0.08}_{-0.07}$) for red and blue galaxies, respectively. This is shallower than the RCS2 results, possibly because of multiple differences in galaxy selection, as discussed in detail in Velander *et al.* (2014); however, their results are largely in agreement with the results from the SDSS. Comparison with COSMOS results is somewhat more complex, given that the best-fitting stellar vs. halo mass relationship can only be turned into an average halo mass within each bin when considering the scatter in stellar mass at fixed halo mass. We will return to this point shortly.

Han *et al.* (2014) carried out a maximum-likelihood weak lensing analysis of optically-selected galaxy groups in the Galaxy and Mass Assembly (GAMA) survey, using SDSS galaxies as background sources. While the focus of the paper is on groups of galaxies, results are also shown for a set of "groups" with $N_{\rm gal} = 1$, i.e., individual galaxies split by stellar mass (not type or morphology). The trends for the average halo mass as a function of stellar mass were largely consistent with those from previous surveys. One point that was highlighted in both Leauthaud *et al.* (2011) and Han *et al.* (2014), and that is quite relevant to the comparison of results from e.g. Velander *et al.* (2014) against those from Leauthaud *et al.* (2012a), is that the average halo mass at fixed stellar mass is very sensitive to the lognormal dispersion in M_* at fixed $M_{\rm halo}$ (even more so than on the median M_* at fixed $M_{\rm halo}$). Thus, comparison of results of contraining M_* as a function of $M_{\rm halo}$ against measurements of $\langle M_{\rm halo} \rangle$ at fixed M_* is highly non-trivial. For the GAMA galaxies, Han *et al.* (2014) found $\sigma_{\log M_*} \sim 0.15$, consistent with the COSMOS results within the errors. However, since $\sigma_{\log M_*}$ includes not just intrinsic scatter but also measurement error (which could depend on how the stellar mass are estimated and

† The errors are similar to those in COSMOS at low stellar mass, however in the case of the COSMOS analysis, the clustering and stellar mass function plays a significant role in the halo model constraints at low stellar mass.

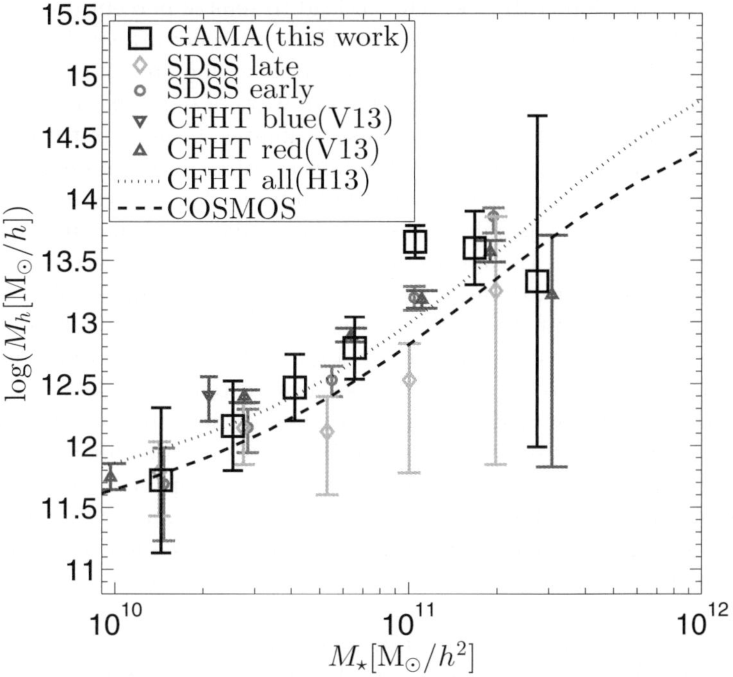

Figure 1. The central halo mass-stellar mass relation measured from weak lensing, from several works described in this section. The quantity that is plotted is the average logarithmic halo mass at fixed stellar mass. The COSMOS measurement, originally provided as a fitted $\langle\log(M_\star)|M_h\rangle$ relation, was converted to the plotted quantity by taking into account the scatter in mean stellar mass at fixed halo mass. The same conversion is done for the results from Hudson *et al.* (2013). It is important to keep in mind that $\langle M_{\rm halo}\rangle(M_*)$ depends not just on the mean relationship between these quantities, but also on both their intrinsic scatter and the measurement error in the M_*. For example, surveys with stellar masses based purely on photometric redshifts, which are an additional source of non-negligible scatter, should have lower average halo masses at fixed stellar mass. Figure used with permission from Han *et al.* (2014).

whether the dataset has spectroscopic or photometric redshifts), it is not necessarily the case that we even expect similar results from all surveys.

To summarize, in Fig. 1 we show figure 9 from Han *et al.* (2014), which compares the results for the average halo mass in stellar mass bins (averaged over all galaxy types) from several different papers.

4. Future challenges

Aside from the basic observational challenges in weak lensing measurements, which are summarized in e.g. Weinberg *et al.* (2013), here I discuss some of the challenges in how to make and interpret weak lensing measurements of dark matter halo masses.

The first issue is in defining the observational quantity to be used as a mass tracer, typically either stellar mass and luminosity. For galaxy samples with spectra, the presence of a redshift can be useful in making a robust estimate of these quantities by fitting the SED to templates. However, for lens samples with imaging data alone, photometric redshifts uncertainty becomes uncertainty in the stellar mass and luminosity. The process of carrying out these fits to interpret the SED in terms of stellar mass and luminosity has other sources of uncertainty. The most commonly considered uncertainty is due to the stellar initial mass function (IMF), which can lead to tens of percent differences in stellar mass estimates. However, even controlling for IMF differences, Leauthaud *et al.* (2012b) argue that differences in stellar mass estimates due to the stellar population synthesis model, form chosen for star formation history, and ways of handling dust attenuation can lead to $\sim 45\%$ uncertainties in the stellar mass estimates. Finally, in the estimate of stellar mass from the mass-to-light ratio and luminosity, we cannot ignore uncertainties that may arise due to the flux estimate that goes into luminosity. This estimate often involves fitting the surface brightness profile of the galaxies, which itself has many pitfalls in the choice of models to use (e.g., Bernardi *et al.* 2014) and systematic errors due to sky level misestimates near bright foregrounds. These sources of uncertainty make comparison between measurements from different surveys with different analysis techniques necessarily difficult, and in future, we will need to control for such differences quite carefully.

Another issue, which is relevant for studies that attempt to define type-dependent relationships between stellar and halo mass, relates to how the galaxies were split into different types. Morphological estimators can be noisy and difficult to apply to limited-resolution imaging from the ground. Estimators based on spectra, such as the strength of the 4000Å break, are limited to the special cases where spectra for lens objects are available. Use of rest-frame colors from imaging data can couple photometric redshift errors to the quantity used for type classification, and can also be fooled by dust extinction in edge-on spirals. It is therefore not apparent to what degree we can compare type-dependent relationships that are derived in different ways. It is interesting that Hudson *et al.* (2013) and Velander *et al.* (2014) found using CFHTLenS that for certain stellar mass bins, their estimates of the halo vs. stellar mass relation differ for certain galaxy types. Naively, since they are using the same source catalog and photometric redshifts, one would expect the noise in their measurements to be correlated, which should result in their agreeing at better than the 1σ level. However, the two works differ in their choice of type separator (color vs. spectral type, respectively) and their way of binning the galaxies by stellar mass (directly vs. using luminosity and then getting an average stellar mass for each luminosity bin). They also have some differences in how the mass modeling of the lensing signals was carried out. Differences in how the data were split into samples and modeled must be responsible for the difference between these results.

The final challenges have to do with how halo mass is modeled. The choice of whether to use isolated galaxies (which do not require halo modeling) or not could lead to the selection of unfair samples compared to use of all galaxies regardless of environment. At the same time, the halo model that can be used to interpret the lensing signal for a sample of mixed central and satellite galaxies has its own limitations: for example, it does not include assembly bias (wherein galaxy properties depend not just on halo mass but some other parameter like age; see Zentner *et al.* 2014 regarding its effects on halo model analyses, and Hearin *et al.* 2014 for an analysis that attempts to account for assembly bias). It also relies on N-body based quantities like the halo mass function and concentration vs. mass relation, which means that if baryonic effects are important (Duffy *et al.* 2010; Cui *et al.* 2014) then the halo model predictions will be incorrect and therefore so will the estimated masses. Another factor that is relevant to the studies that already use halo models is the choice of how much complexity to allow. For example, Mandelbaum *et al.* (2006), van Uitert *et al.* (2011), and Velander *et al.* (2014) all used simple halo models with many parameters fixed and with simple corrections for halo mass vs. observable scatter; Leauthaud *et al.* (2011) and Tinker *et al.* (2013) used a much more complex halo model with more freedom and fewer fixed parameters, but they also needed to include much more data (stellar mass function and clustering) in order to get strong parameter constraints, which means that systematic uncertainties in abundances become more relevant than in the lensing-only studies. It is clear that some of these nuisance parameters are important, and in order to compare results we must understand the impact of choices about which nuisance parameters are held fixed and which are free.

In short, the above complications can complicate a comparison between the different measurements described in Section 3. In the near future, larger lensing surveys are going to make even higher S/N measurements of galaxy-galaxy lensing, which could lead to a better understanding of the halo vs. dark matter connection from weak lensing. At the same time, with the better data will come a need for greater understanding of how these challenges are affecting the results.

5. Conclusions

Because of its sensitivity to all types of matter (stars, gas, and dark matter) and its insensitivity to their dynamical state, weak gravitational lensing has emerged as a powerful tool to study the connection between galaxies and their host dark matter halos in the past decade, with results from multiple surveys, the most recent of which have achieved statistical errors in the 5% regime. While there are a number of challenges to understand in the modeling that is used to go from measured lensing signals to constraints on masses, as described in §4, there is every reason to believe that near-term surveys like Hyper Suprime-Cam (HSC†, Miyazaki *et al.* 2006), the Dark Energy Survey (DES‡, The Dark Energy Survey Collaboration 2005), and the KIlo-Degree Survey (KIDS¶) will continue to improve on what we have already learned, and we will learn yet more from the even more ambitious programs that are planned for the coming decade, including the Large Synoptic Survey Telescope (LSST‖, LSST Science Collaborations *et al.* 2009), Euclid†† (Laureijs *et al.* 2011), and the Wide-Field Infrared Survey Telescope (WFIRST AFTA project‡‡, Spergel *et al.* 2013).

Acknowledgements

I am grateful for the support of the Alfred P. Sloan Foundation, which supported my work on this review. I would like to thank Jiaxin Han for permitting me to use figure 9 from his paper, and Alexie Leauthaud and Ying Zu for useful discussions.

References

Bernardi, M., Meert, A., Vikram, V., Huertas-Company, M., Mei, S., Shankar, F., & Sheth, R. K., 2014, *MNRAS*, 443, 874

Cacciato, M., van Uitert, E., & Hoekstra, H., 2014, *MNRAS*, 437, 377

Cooray, A. & Sheth, R., 2002, *Phys. Rep.*, 372, 1

Cui, W., Borgani, S., & Murante, G., 2014, *MNRAS*, 441, 1769

Duffy, A. R., Schaye, J., Kay, S. T., Dalla Vecchia, C., Battye, R. A., & Booth, C. M., 2010, *MNRAS*, 405, 2161

Dutton, A. A., Conroy, C., van den Bosch, F. C., Prada, F., & More, S., 2010, *MNRAS*, 407, 2

Erben, T., Hildebrandt, H., Miller, L., van Waerbeke, L., Heymans, C., Hoekstra, H., Kitching,

† `http://www.naoj.org/Projects/HSC/index.html`
‡ `https://www.darkenergysurvey.org/`
¶ `http://www.astro-wise.org/projects/KIDS/`
‖ `http://www.lsst.org/lsst`
†† `http://sci.esa.int/euclid`, `http://www.euclid-ec.org`
‡‡ `http://wfirst.gsfc.nasa.gov/`

T. D., Mellier, Y., Benjamin, J., Blake, C., Bonnett, C., Cordes, O., Coupon, J., Fu, L., Gavazzi, R., Gillis, B., Grocutt, E., Gwyn, S. D. J., Holhjem, K., Hudson, M. J., Kilbinger, M., Kuijken, K., Milkeraitis, M., Rowe, B. T. P., Schrabback, T., Semboloni, E., Simon, P., Smit, M., Toader, O., Vafaei, S., van Uitert, E., & Velander, M., 2013, *MNRAS*, 433, 2545

Gavazzi, R., Treu, T., Rhodes, J. D., Koopmans, L. V. E., Bolton, A. S., Burles, S., Massey, R. J., & Moustakas, L. A., 2007, *ApJ*, 667, 176

Gilbank, D. G., Gladders, M. D., Yee, H. K. C., & Hsieh, B. C., 2011, *AJ*, 141, 94

Gladders, M. D. & Yee, H. K. C., 2005, *ApJS*, 157, 1

Guzik, J. & Seljak, U., 2001, *MNRAS*, 321, 439

Han, J., Eke, V. R., Frenk, C. S., Mandelbaum, R., Norberg, P., Schneider, M. D., Peacock, J. A., Jing, Y., Baldry, I., Bland-Hawthorn, J., Brough, S., Brown, M. J. I., Liske, J., Loveday, J., & Robotham, A. S. G., 2014, preprint (arXiv:1404.6828)

Hearin, A. P., Watson, D. F., Becker, M. R., Reyes, R., Berlind, A. A., & Zentner, A. R., 2014, *MNRAS*, 444, 729

Heymans, C., Bell, E. F., Rix, H.-W., Barden, M., Borch, A., Caldwell, J. A. R., McIntosh, D. H., Meisenheimer, K., Peng, C. Y., Wolf, C., Beckwith, S. V. W., Hä,uß,ler B., Jahnke, K., Jogee, S., Sá,nchez S. F., Somerville, R., & Wisotzki, L., 2006, *MNRAS*, 371, L60

Heymans, C., Van Waerbeke, L., Miller, L., Erben, T., Hildebrandt, H., Hoekstra, H., Kitching, T. D., Mellier, Y., Simon, P., Bonnett, C., Coupon, J., Fu, L., Harnois Dé,raps J., Hudson, M. J., Kilbinger, M., Kuijken, K., Rowe, B., Schrabback, T., Semboloni, E., van Uitert, E., Vafaei, S., & Velander, M., 2012, *MNRAS*, 427, 146

Hoekstra, H., Bartelmann, M., Dahle, H., Israel, H., Limousin, M., & Meneghetti, M., 2013, *Space Sci. Rev.*, 177, 75

Hoekstra, H., Hsieh, B. C., Yee, H. K. C., Lin, H., & Gladders, M. D., 2005, *ApJ*, 635, 73

Hoekstra, H. & Jain, B., 2008, *Annual Review of Nuclear and Particle Science*, 58, 99

Hudson, M. J., Gillis, B. R., Coupon, J., Hildebrandt, H., Erben, T., Heymans, C., Hoekstra, H., Kitching, T. D., Mellier, Y., Miller, L., Van Waerbeke, L., Bonnett, C., Fu, L., Kuijken, K., Rowe, B., Schrabback, T., Semboloni, E., van Uitert, E., & Velander, M., 2013, preprint (arXiv:1310.6784)

Koekemoer, A. M., Aussel, H., Calzetti, D., Capak, P., Giavalisco, M., Kneib, J., Leauthaud, A., Le Fèvre, O., McCracken, H. J., Massey, R., Mobasher, B., Rhodes, J., Scoville, N., & Shopbell, P. L., 2007, *ApJS*, 172, 196

Laureijs, R., Amiaux, J., Arduini, S., Auguè,res J. ., Brinchmann, J., Cole, R., Cropper, M., Dabin, C., Duvet, L., & Ealet, A., *et al.* 2011, preprint (arXiv:1110.3193)

Leauthaud, A., Tinker, J., Behroozi, P. S., Busha, M. T., & Wechsler, R. H., 2011, *ApJ*, 738, 45

Leauthaud, A., Tinker, J., Bundy, K., Behroozi, P. S., Massey, R., Rhodes, J., George, M. R., Kneib, J.-P., Benson, A., Wechsler, R. H., Busha, M. T., Capak, P., Cortês, M., Ilbert, O., Koekemoer, A. M., Le Fèvre, O., Lilly, S., McCracken, H. J., Salvato, M., Schrabback, T., Scoville, N., Smith, T., & Taylor, J. E., 2012a, *ApJ*, 744, 159

Leauthaud, A., George, M. R., Behroozi, P. S., Bundy, K., Tinker, J., Wechsler, R. H., Conroy, C., Finoguenov, A., & Tanaka, M., 2012b, *ApJ*, 746, 95

LSST Science Collaborations, Abell, P. A., Allison, J., Anderson, S. F., Andrew, J. R., Angel, J. R. P., Armus, L., Arnett, D., Asztalos, S. J., & Axelrod, T. S., *et al.* 2009, preprint (arXiv:0912.0201)

Mandelbaum, R., Li, C., Kauffmann, G., & White, S. D. M., 2009, *MNRAS*, 393, 377

Mandelbaum, R., Seljak, U., & Hirata, C. M., 2008, *J. Cosmology Astropart. Phys.*, 8, 6

Mandelbaum, R., Seljak, U., Kauffmann, G., Hirata, C. M., & Brinkmann, J., 2006, *MNRAS*, 368, 715

Mandelbaum, R., Tasitsiomi, A., Seljak, U., Kravtsov, A. V., & Wechsler, R. H., 2005, *MNRAS*, 362, 1451

Miyazaki, S., Komiyama, Y., Nakaya, H., Doi, Y., Furusawa, H., Gillingham, P., Kamata, Y., Takeshi, K., & Nariai, K., 2006, in *Society of Photo-Optical Instrumentation Engineers (SPIE) Conference Series Vol. 6269 of Society of Photo-Optical Instrumentation Engineers (SPIE) Conference Series*, HyperSuprime: project overview

Navarro, J. F., Frenk, C. S., & White, S. D. M., 1996, *ApJ*, 462, 563

Reyes, R., Mandelbaum, R., Gunn, J. E., Nakajima, R., Seljak, U., & Hirata, C. M., 2012, *MNRAS*, 425, 2610

Schulz, A. E., Mandelbaum, R., & Padmanabhan, N., 2010, *MNRAS*, 408, 1463

Scoville, N., Abraham, R. G., Aussel, H., Barnes, J. E., Benson, A., Blain, A. W., Calzetti, D., Comastri, A., Capak, P., Carilli, C., Carlstrom, J. E., Carollo, C. M., Colbert, J., Daddi, E., Ellis, R. S., Elvis, M., Ewald, S. P., Fall, M., Franceschini, A., Giavalisco, M., Green, W., Griffiths, R. E., Guzzo, L., Hasinger, G., Impey, C., Kneib, J., Koda, J., Koekemoer, A., Lefevre, O., Lilly, S., Liu, C. T., McCracken, H. J., Massey, R., Mellier, Y., Miyazaki, S., Mobasher, B., Mould, J., Norman, C., Refregier, A., Renzini, A., Rhodes, J., Rich, M., Sanders, D. B., Schiminovich, D., Schinnerer, E., Scodeggio, M., Sheth, K., Shopbell, P. L., Taniguchi, Y., Tyson, N. D., Urry, C. M., Van Waerbeke, L., Vettolani, P., White, S. D. M., & Yan, L., 2007a, *ApJS*, 172, 38

Scoville, N., Aussel, H., Brusa, M., Capak, P., Carollo, C. M., Elvis, M., Giavalisco, M., Guzzo, L., Hasinger, G., Impey, C., Kneib, J., LeFevre, O., Lilly, S. J., Mobasher, B., Renzini, A., Rich, R. M., Sanders, D. B., Schinnerer, E., Schminovich, D., Shopbell, P., Taniguchi, Y., & Tyson, N. D., 2007b, *ApJS*, 172, 1

Seljak, U., 2000, *MNRAS*, 318, 203

Spergel, D., Gehrels, N., Breckinridge, J., Donahue, M., Dressler, A., Gaudi, B. S., Greene, T., Guyon, O., Hirata, C., Kalirai, J., Kasdin, N. J., Moos, W., Perlmutter, S., Postman, M., Rauscher, B., Rhodes, J., Wang, Y., Weinberg, D., Centrella, J., Traub, W., Baltay, C., Colbert, J., Bennett, D., Kiessling, A., Macintosh, B., Merten, J., Mortonson, M., Penny, M., Rozo, E., Savransky, D., Stapelfeldt, K., Zu, Y., Baker, C., Cheng, E., Content, D., Dooley, J., Foote, M., Goullioud, R., Grady, K., Jackson, C., Kruk, J., Levine, M., Melton, M., Peddie, C., Ruffa, J., & Shaklan, S., 2013, preprint (arXiv:1305.5422)

The Dark Energy Survey Collaboration, 2005, ArXiv Astrophysics e-prints

Tinker, J. L., Leauthaud, A., Bundy, K., George, M. R., Behroozi, P., Massey, R., Rhodes, J., & Wechsler, R. H., 2013, *ApJ*, 778, 93

van den Bosch, F. C., More, S., Cacciato, M., Mo, H., & Yang, X., 2013, *MNRAS*, 430, 725

van Uitert, E., Hoekstra, H., Velander, M., Gilbank, D. G., Gladders, M. D., & Yee, H. K. C., 2011, *A&A*, 534, A14

Velander, M., van Uitert, E., Hoekstra, H., Coupon, J., Erben, T., Heymans, C., Hildebrandt, H., Kitching, T. D., Mellier, Y., Miller, L., Van Waerbeke, L., Bonnett, C., Fu, L., Giodini, S., Hudson, M. J., Kuijken, K., Rowe, B., Schrabback, T., & Semboloni, E., 2014, *MNRAS*, 437, 2111

Weinberg, D. H., Mortonson, M. J., Eisenstein, D. J., Hirata, C., Riess, A. G., & Rozo, E., 2013, *Phys. Rep.*, 530, 87

York, D. G., Adelman, J., Anderson, Jr. J. E., Anderson, S. F., Annis, J., Bahcall, N. A., Bakken, J. A., Barkhouser, R., Bastian, S., Berman, E., Boroski, W. N., Bracker, S., Briegel, C., Briggs, J. W., Brinkmann, J., Brunner, R., Burles, S., Carey, L., Carr, M. A., Castander, F. J., Chen, B., Colestock, P. L., Connolly, A. J., Crocker, J. H., Csabai, I., Czarapata, P. C., Davis, J. E., Doi, M., Dombeck, T., Eisenstein, D., Ellman, N., Elms, B. R., Evans, M. L., Fan, X., Federwitz, G. R., Fiscelli, L., Friedman, S., Frieman, J. A., Fukugita, M., Gillespie, B., Gunn, J. E., Gurbani, V. K., de Haas E., Haldeman, M., Harris, F. H., Hayes, J., Heckman, T. M., Hennessy, G. S., Hindsley, R. B., Holm, S., Holmgren, D. J., Huang, C.-h., Hull, C., Husby, D., Ichikawa, S.-I., Ichikawa, T., Ivezić, Ž,., Kent, S., Kim, R. S. J., Kinney, E., Klaene, M., Kleinman, A. N., Kleinman, S., Knapp, G. R., Korienek, J., Kron, R. G., Kunszt, P. Z., Lamb, D. Q., Lee, B., Leger, R. F., Limmongkol, S., Lindenmeyer, C., Long, D. C., Loomis, C., Loveday, J., Lucinio, R., Lupton, R. H., MacKinnon, B., Mannery, E. J., Mantsch, P. M., Margon, B., McGehee, P., McKay, T. A., Meiksin, A., Merelli, A., Monet, D. G., Munn, J. A., Narayanan, V. K., Nash, T., Neilsen, E., Neswold, R., Newberg, H. J., Nichol, R. C., Nicinski, T., Nonino, M., Okada, N., Okamura, S., Ostriker, J. P., Owen, R., Pauls, A. G., Peoples, J., Peterson, R. L., Petravick, D., Pier, J. R., Pope, A., Pordes, R., Prosapio, A., Rechenmacher, R., Quinn, T. R., Richards, G. T., Richmond, M. W., Rivetta, C. H., Rockosi, C. M., Ruthmansdorfer, K., Sandford, D., Schlegel, D. J., Schneider, D. P., Sekiguchi, M., Sergey, G., Shimasaku, K., Siegmund, W. A., Smee, S., Smith, J. A., Snedden, S., Stone, R., Stoughton, C., Strauss, M. A., Stubbs, C., SubbaRao, M., Szalay, A. S., Szapudi, I., Szokoly, G. P., Thakar, A. R., Tremonti, C., Tucker, D. L., Uomoto, A., Vanden Berk, D., Vogeley, M. S., Waddell, P., Wang, S.-i., Watanabe, M., Weinberg, D. H., Yanny, B., Yasuda, N., & SDSS Collaboration 2000, *AJ*, 120, 1579

Zentner, A. R., Hearin, A. P., & van den Bosch, F. C., 2014, *MNRAS*, 443, 3044

Galaxy Masses as Constraints of Formation Models
Proceedings IAU Symposium No. 311, 2014
M. Cappellari & S. Courteau, eds.

doi:10.1017/S1743921315003464

Stellar masses calibrated with micro-lensed quasars

Paul L. Schechter[1], **Jeffrey A. Blackburne**[2], **David Pooley**[3] **and Joachim Wambsganss**[4]

[1]MIT Kavli Institute Cambridge, MA 02139, USA
email: schech@mit.edu
[2] Areté Associates, Northridge, CA 91324
[3] Sam Houston State University, Huntsville, TX 77341
[4] Zentrum für Astronomie der Universität Heidelberg, Germany

Abstract. We measure the stellar mass surface densities of early type galaxies by observing the micro-lensing of macro-lensed quasars caused by individual stars, including stellar remnants, brown dwarfs and red dwarfs too faint to produce photometric or spectroscopic signatures. Our method measures the *graininess* of the gravitational potential, in contrast to methods that decompose a smooth total gravitational potential into two smooth components, one stellar and one dark. We find the median likelihood value for the calibration factor $\mathcal{F}$ by which Salpeter stellar masses (with a low mass cutoff of $0.1 M_\odot$) must be multiplied is 1.23, with a one sigma confidence range of $0.77 < \mathcal{F} < 2.10$.

Keywords. galaxies: stellar content, gravitational lensing

1. Introduction

Stellar masses for early type galaxies are almost always determined by one of two methods: either they are estimated from spectra (and sometimes only broad band colors) or they are deduced by subtracting the contribution of an assumed dark matter component from a combined mass inferred from kinematic (and sometimes macro-lensing) measurements. Multiple examples of both methods can be found in the present volume. Both methods have shortcomings.

Here we use a third method: determining the stellar mass surface density of an early type galaxy from brightness fluctuations of the four images of a background quasar that is both multiply-imaged ("macro-lensed") by the galaxy and "micro-lensed" by the individual stars in that galaxy (Schechter & Wambsganss 2004; Kochanek 2004). This method, in contrast to spectral methods, is sensitive to stellar mass near and below the hydrogen burning limit, as well as to the mass in stellar remnants. And where dark matter subtraction methods make asumptions about the dark matter profiles, the gravitational micro-lensing technique makes only an assumption about the combined gravitational potential, one that has been subjected to extensive observational verification.

Ideally one would observe a single system long enough to see a great many fluctuations and infer an accurate stellar surface density. But the timescale for micro-lensing variations is of order ten years for a lens at redshift $z \sim 0.5$ (Mosquera & Kochanek 2011), and observations of even four quasar images at a single epoch give only broad constraints on the mass surface density. So instead we observe ten systems at a single epoch and combine results.

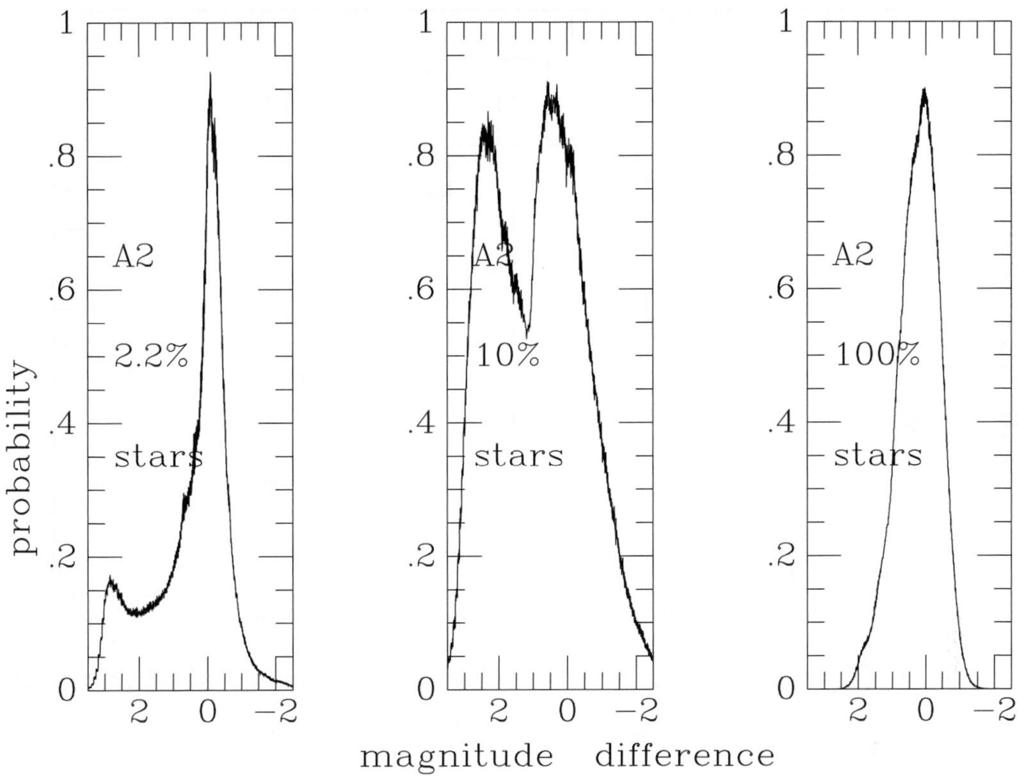

Figure 1. Probability distribution for the ratio of observed to macro-model flux (expressed as a magnitude difference) for the *A*2 image of the quadruple lens PG 1115+080 for three different stellar mass fractions. The different shapes of the distributions permit determination of the stellar mass fraction.

2. Calibrating the stellar mass fundamental plane

Our approach is to use the stellar mass fundamental plane (Hyde & Bernardi 2009) to predict the stellar surface mass densities at the positions of our quasar images, which cause differences between the observed quasar fluxes and those predicted by a macro-model. We then adjust the stellar masses by a multiplicative constant, $\mathcal{F}$. The constant is varied and the likelihood of the observed quasar X-ray fluxes is computed for each value of $\mathcal{F}$.

Our stellar mass fundamental plane was constructed using data from Auger *et al.* (2010) and Sonnenfeld *et al.* (2013), who used a Salpeter initial mass function (IMF) with a low mass cutoff of $0.1M_{\odot}$ to compute stellar masses for lensing galaxies at $z \sim 0.2$ and $z \sim 0.5$ respectively. The median likelihood value of the factor by which these masses must be multiplied is $\mathcal{F} = 1.23$ with a 68% confidence interval of $0.77 < \mathcal{F} < 2.10$. The range is the result of small sample size.

3. Salient features of the analysis

• We take the quasar to be point-like at X-ray wavelengths (relative to the Einstein rings of the micro-lensing stars) and use Chandra fluxes taken from Pooley *et al.* (2007). Blackburne *et al.* (2011) have shown that the optical and near-IR emitting regions are comparable in projected size to the stellar Einstein rings, rendering optical wavelengths less well suited to the present analysis.

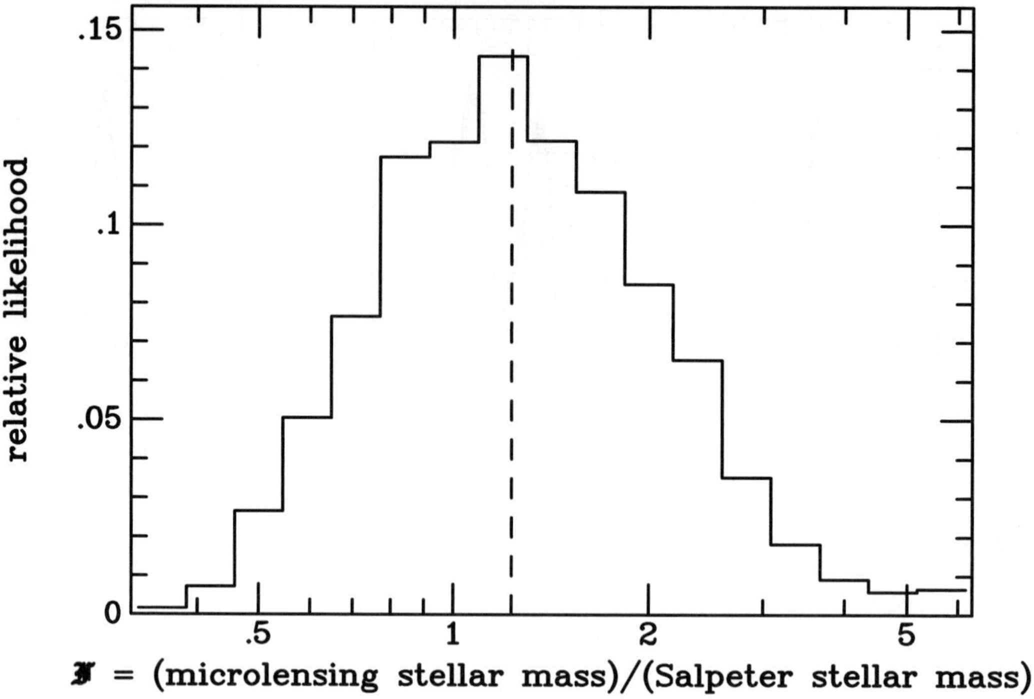

Figure 2. The likelihood histogram for the calibration factor $\mathcal{F}$ by which the stellar mass fundamental plane computed using a Salpeter IMF must be multiplied to produce the observed X-ray flux ratio anomalies.

- We use a fundamental plane in which a "proxy" velocity dispersion computed from the radius of the Einstein ring of the macro-model is substituted for stellar dispersions. We find that for the Auger *et al.* (2010) sample this gives considerably less scatter than using proper stellar velocity dispersions in constructing the fundamental plane.
- We model the lenses as singular isothermal ellipsoids (ellipticity and orientation identical to that of the stellar light) with an external shear.
- We have analyzed the sensistivity of our result to a number of different possible sources of systematic error. We find them to be small compared to the statistical confidence interval. In particular we find that milli-lensing by dark matter substructure has very little effect on our result.
- Our result is most sensitive to the measured de Vaucouleurs radii of the lensing galaxies. The published values fall into two groups, based on the software used, which deviate systematically from each other; we adopt the geometric mean.
- An alternative analysis that takes surface mass density and effective radius to be functions of the proxy velocity dispersion – a "fundamental line" – circumvents the systematic uncertainties in the measured effective radii and yields a calibration factor $\mathcal{F}$ that differs by 7%.
- The X-ray fluxes for the images of RX J0911+0551 pull our calibration factor $\mathcal{F}$ to higher values. The macro-models predict that image b, a minimum of the light travel time, should be more than five times brighter than image d, also a minimum, but is observed to be only 25% brighter.
- For a galaxy with a proxy velocity dispersion of 266 km/s and an effective radius of 6.17 kpc, average for the Auger *et al.* (2010) sample, the calibrated fractional stellar

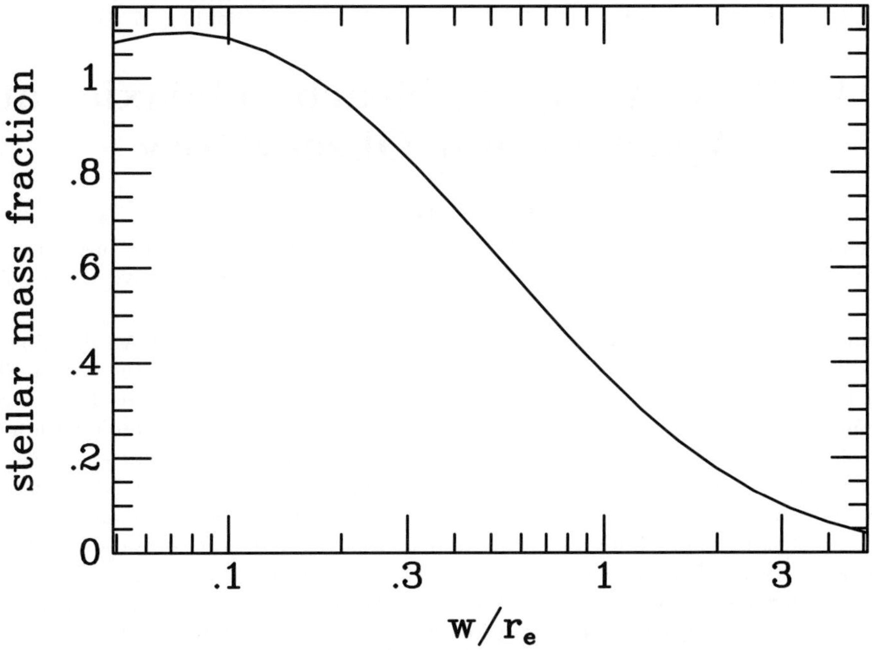

Figure 3. The stellar surface mass density as a fraction of the total for the typical lensing galaxy, obtained by applying the calibration factor $\mathcal{F}$ to the stellar mass fundamental plane derived from the SLACS + SL2S samples. The peak at $w/r_e = 0.074$ (where w is the circularized radius) occurs where the slope of the de Vaucouleurs profile is equal to that of the singular isothermal.

surface mass density peaks at a value of 1.10 at $0.074r_e$, as shown in Figure 3. At $1.5r_e$, the typical Einstein radius for our lens systems, the stellar mass fraction is 0.25.

The work summarized here is the culmination of an effort begun by Schechter & Wambsganss (2004) and described in the proceedings of IAU Symposium 220. A more thorough description of the present analysis can be found in Schechter *et al.* (2014)

Acknowledgements

The authors gratefully acknowledge the very substantial contributions of Saul A. Rappaport. This work was supported in part by the US National Science Foundation under grants AST02-06010 and AST06-07601 and by NASA under Chandra grant G07-8099. The first author thanks the members of the organizing committee for their good efforts.

References

Auger, M. W., Treu, T., Bolton, A. S., *et al.* 2010, *ApJ*, 724, 511
Bate, N. F., Floyd, D. J. E., Webster, R. L., & Wyithe, J. S. B. 2011, *ApJ*, 731, 71
Blackburne, J. A., Pooley, D., Rappaport, S., & Schechter, P. L. 2011, *ApJ*, 729, 34
Hyde, J. B. & Bernardi, M. 2009, *MNRAS*, 396, 1171
Kochanek, C. S. 2004, *ApJ*, 605, 58
Mosquera, A. M. & Kochanek, C. S. 2011, *ApJ*, 738, 96
Pooley, D., Blackburne, J. A., Rappaport, S., & Schechter, P. L. 2007, *ApJ*, 661, 19
Schechter, P. L., Pooley, D., Blackburne, J. A., & Wambsganss, J. 2014, *arXiv:1405.0038*
Schechter, P. L. & Wambsganss, J. 2004, in *IAU Symposium, Vol. 220, Dark Matter in Galaxies*, ed. S. Ryder, D. Pisano, M. Walker, & K. Freeman, 103.
Sonnenfeld, A., Gavazzi, R., Suyu, S. H., Treu, T., & Marshall, P. J. 2013, *ApJ*, 777, 97

Galaxy Masses as Constraints of Formation Models
Proceedings IAU Symposium No. 311, 2014
M. Cappellari & S. Courteau, eds.

doi:10.1017/S1743921315003476

MaNGA: Mapping Nearby Galaxies at Apache Point Observatory

Kevin Bundy

Kavli Institute for the Physics and Mathematics of the Universe, Todai Institutes for Advanced Study, the University of Tokyo, Kashiwa, Japan 277-8583 (Kavli IPMU, WPI)

Abstract. I describe a new integral field spectroscopic survey called MaNGA (Mapping Nearby Galaxies at Apache Point Observatory). One of three core programs in the 6-year SDSS-IV project† that began on July 1st, 2014, MaNGA will deploy 17 fiber-bundle IFUs across the Sloan 2.5m Telescope's 3 degree field-of-view, targeting a mass-selected sample with a median redshift of 0.03, typical spatial resolution of 1-2 kpc, and a per-fiber signal-to-noise ratio of 4-8 in the outskirts of target galaxies. For each galaxy in the sample, MaNGA will provide maps and measured gradients of the composition and dynamics of both stars and gas. I discuss early results that highlight MaNGA's potential to shed light on the ionization and chemical enrichment of gas in galaxies, spatial patterns in their star formation histories, and the internal makeup of stellar populations. MaNGA's unprecedented data set will not only provide powerful new insight on galaxy formation and evolution but will serve as a valuable benchmark for future high-z observations from large telescopes and space-based facilities.

Keywords. galaxies: evolution - galaxies: formation

1. Introduction

Our understanding of galaxy formation has significantly advanced in the last decade thanks to SDSS surveys at low redshift, a variety of higher redshift surveys and observations, and improvements in simulating non-linear structure formation. While at the present day the largest dark matter halos continue to grow from the "bottom-up" by merging with smaller halos, the galaxies living in these halos experienced peak star formation rates at $z \sim 1$–3. From this time on, activity steadily declined while galaxies continued to grow via mergers and star formation fueled by the accretion of fresh gas. Eventually, star formation was somehow shut down. This quenching process, a kind of "galaxy death," increasingly affects lower mass systems with time and may soon be the fate of our own Milky Way. Present-day galaxies less massive than the Milky Way formed their stars continuously over all cosmic epochs. In fact, the majority of the stars in galactic disks formed in the last 7-9 billion years.

In spite of significant progress in delineating this basic picture, important pieces of the puzzle remain to be understood. Very importantly, we still seek *the physical origin* of the patterns of evolution described above. Gravitational collapse is only one of many processes at work in the formation of a galaxy. Gas dissipation and infall, star formation, heating and ejection of gas by supernova explosions, galaxy merging, the redistribution of material due to internal processes, and associated growth of central black holes are all key processes that remain poorly understood.

One limitation has been the fact that a single spectroscopic measurement at a galaxy's center cannot capture the rich physics responsible for its internal structure and dynamics. Two-dimensional, integral-field spectroscopy (IFS) is needed and recent efforts such as

† `www.sdss.org`

SAURON (de Zeeuw *et al.* 2002), DiskMass (Bershady *et al.* 2010), ATLAS^{3D} (Cappellari *et al.* 2011), and CALIFA (Sánchez *et al.* 2012) have demonstrated enormous promise, although their small (few $\times 100$) and incomplete samples hamper statistical studies. The more recently launched SAMI survey (utilizing the Sydney Australian Astronomical Observatory Multi-object Integral Field Spectrograph, Croom *et al.* 2012) is making major gains using a multiplexed fiber-IFU instrument. MaNGA represents another advance, surveying 10,000 galaxies across a wide dynamic range in M_*, environment, and star formation rate with uniform radial coverage. Especially important is MaNGA's unique wavelength coverage in the near-IR, the location of diagnostics sensitive to the IMF, such as NaI and the CaII triplet (Conroy & van Dokkum 2012), as well as important tracers of thermally-pulsating asymptotic giant branch (TP-AGB) stars and other intermediate-age populations (Maraston 2005). Combined with its sensitivity in the near-UV, MaNGA will open up simultaneous access to a large number of compositional features and dynamic tracers. A comprehensive overview of the MaNGA project is given in Bundy *et al.* (2014).

2. Instrument and Survey Design

The MaNGA instrument couples a novel technology that delivers regular, high-throughput fiber integral field units (IFUs) with the sensitive spectrographs. The instrument and the target selection have been designed and optimized in concert.

MaNGA IFUs: MaNGA's high-performance fiber bundles have "fill factors" of $\sim$56%. To uniformly cover targets in order to reconstruct their light distribution requires requires dithered observations and highly uniform fiber spacing. Through the development of breakthrough technology (see below), MaNGA achieves a packing regularity that is 1–2% (3μm) from the ideal hexagonal configuration. Careful attention to fiber stress, length, and application of anti-reflective coating allows MaNGA to reach lab-measured 95% throughput across all fibers in our bundles. Our fabrication technology scales easily to yield a range of IFU sizes from 19 to 127 fibers with identical performance optimized for our galaxy sample.

Spectroscopy: MaNGA makes use of the high-sensitivity BOSS spectrographs (Smee *et al.* 2013), each of which features a red and blue arm for continuous wavelength coverage from the near-UV into the near-IR. MaNGA packs 1423 fibers into the spectrographs, enabling input from 17 science IFU bundles, twelve 7-fiber flux calibration bundles, and 92 sky fibers. With a typical resolution of $R \sim 2000$, MaNGA samples a broad wavelength region, including the region of 750 nm$< \lambda <$ 1000 nm. This expansive and continuous coverage captures a large range of absorption features, from blue indices such as D4000, Ca H+K, and higher-order Balmer lines, through classical optical absorption features such as Hβ, Mgb, and Fe5270/Fe5335, to red, gravity-sensitive features such as the Ca triplet at 860nm. These features encode a great deal of information about star formation histories and the IMF, and can be used to measure element abundance ratios such as [Mg/Fe], [N/Fe], [C/Fe] and [Ca/Fe]. The spectrograph's sensitivity (ϵ) and solid angle (Ω) combined with the Sloan 2.5m Telescope's (Gunn *et al.* 2006) collecting area (A) yield an unparalleled etendue (A$\Omega\epsilon$).

Sample Optimization: Maximizing S/N and spatial resolution demands target galaxies with the lowest possible redshifts. Typical ground-based sampling of the radial extent of small galaxies with a least three resolution elements (assuming FWHM$\sim$ 2.0") requires $z \lesssim 0.03$. However, building large samples with representative numbers of high-mass galaxies requires large volumes and therefore demands redshifts $z > 0.03$. At a median redshift of ~ 0.03, as a result of the cosmic galaxy number density, a sweet spot exists to

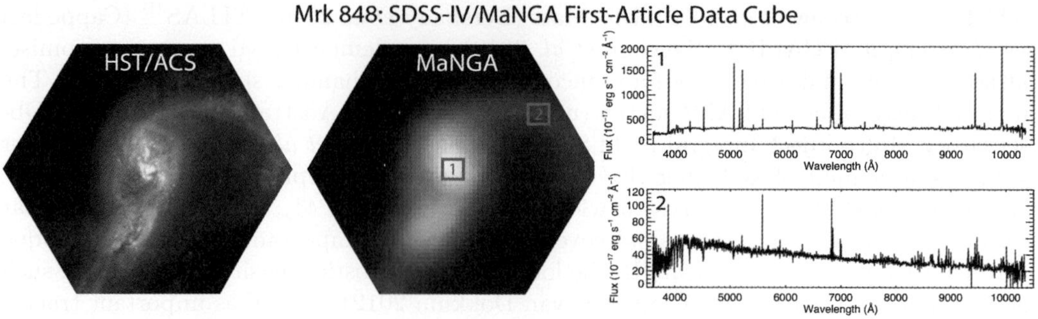

Figure 1. Color HST image of Markarian 848 compared to the reconstructed MaNGA color image for the same galaxy. MaNGA spectra from two different regions are also plotted. Credit: David Law.

gather a sample size of many thousand of galaxies in a survey covering several thousand deg^2. The sky density of such a sample is 3 deg^{-2}, and fortuitously is well matched to the MaNGA IFU complement made possible by Sloan's 7 deg^2 field of view and the BOSS spectrographs.

We have performed an exhaustive optimization of IFU bundle size distribution, yielding 17 IFUs with between 19 and 127 fibers that ensure radial coverage to 1.5 R_e and, for gradient measures requiring the highest S/N ($\gtrsim$ 30 Å^{-1}) via annular binning, a minimum of three resolved annuli (e.g., Δr =2"). On average, MaNGA galaxies are covered by five annular bins that result in sufficient stacked S/N to derive robust stellar population parameters (e.g., age, metallicity, element abundances). For S/N= 8, dithered MaNGA observations provide over 80 unique, radial samples for the average target galaxy (the median MaNGA *physical* spatial resolution is 1–2 kpc). The sample is *volume-limited* as a function of i-band luminosity so that roughly equal numbers of the lowest ($10^9 M_\odot$) and highest mass ($10^{11.5} M_\odot$) galaxies are obtained. We intend to devote 1/3 of the full sample to an exploratory study of the physical properties at 2.5 R_e, a regime never yet probed with significant statistics. MaNGA targets are selected solely from the SDSS-I MAIN sample and are adaptively tiled to probe the full range of central and satellite galaxies in all halos ($\log(M_h/M_\odot)$ = 10.5–15) and local density environments.

3. Early Results

Three studies based on data obtained in 2013 using a prototype of the MaNGA instrument have now been submitted for publication. Belfiore *et al.* study maps of the ionization states of gas in this P-MaNGA sample. Wilkinson *et al.* develop a new spectral fitter to derive maps of the age, metallicity, dust, and surface mass density of the stellar populations. Li *et al.* map diagnostics of recent star formation like D4000 and Hδ to explore constraints on recent fueling and quenching.

Commissioning of the final MaNGA instrumentation began in March 2014 and the first data cubes have been produced. The galaxy, Markarian 848, was targeted on the first MaNGA plate and is shown in Figure 1. A more typical MaNGA galaxy is shown in Figure 2.

Acknowledgements

SDSS is managed by the Astrophysical Research Consortium for the Participating Institutions. These include Carnegie Mellon University, Colorado University, Boulder, Harvard-

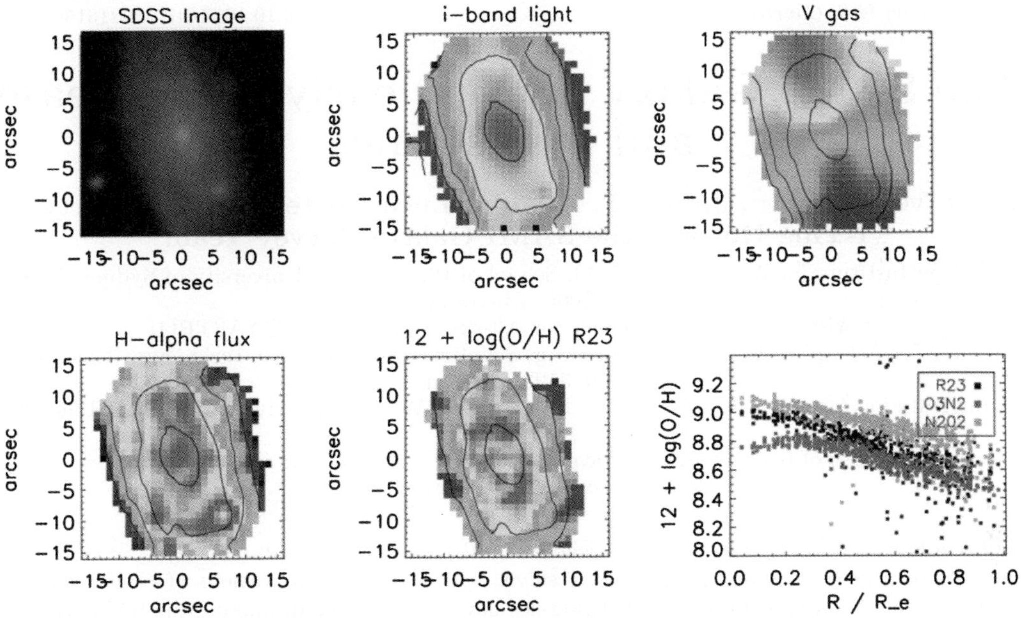

Figure 2. A typical star-forming galaxy as observed during MaNGA commissioning in March 2014. Maps of the *i*-band flux, gas velocity (as measured from Hα), Hαflux, and R23 gas-phase metallicity are shown. The bottom right panel plots gas-phase metallicity gradients using three popular indicators. Individual points represent separate spaxels in the data cube for this galaxy. Credit: Christy Tremonti.

Smithsonian Center for Astrophysics Participation Group, Johns Hopkins University, Kavli Institute for the Physics and Mathematics of the Universe Max-Planck-Institut fuer Astrophysik (MPA Garching), Max-Planck-Institut fuer Extraterrestrische Physik (MPE), Max-Planck-Institut fuer Astronomie (MPIA Heidelberg), National Astronomical Observatories of China, New Mexico State University, New York University, The Ohio State University, Penn State University, Shanghai Astronomical Observatory, United Kingdom Participation Group, University of Portsmouth, University of Utah, University of Wisconsin, and Yale University. This work was supported in part by a World Premier International Research Center Initiative (WPI Initiative), MEXT, Japan.

References

Bershady, M. A., Verheijen, M. A. W., Swaters, R. A., Andersen, D. R., Westfall, K. B., & Martinsson, T. 2010, *ApJ*, 716, 198

Cappellari, M. *et al.* 2011, *MNRAS*, 413, 813

Conroy, C. & van Dokkum, P. 2012, *ApJ*, 747, 69

Croom, S. M. *et al.* 2012, *MNRAS*, 421, 872

de Zeeuw, P. T. *et al.* 2002, *MNRAS*, 329, 513

Gunn, J. E. *et al.* 2006, *AJ*, 131, 2332

Maraston, C. 2005, *MNRAS*, 362, 799

Sánchez, S. F. *et al.* 2012, *A&A*, 538, A8

Smee, S. A., Gunn, J. E., Uomoto, A., *et al.* 2013, *AJ*, 146, 32

Galaxy Masses as Constraints of Formation Models
Proceedings IAU Symposium No. 311, 2014
M. Cappellari & S. Courteau, eds.

doi:10.1017/S1743921315003488

The SAMI Galaxy Survey: early data release and first science

Scott M. Croom[1,2], James T. Allen[1,2], Luca Cortese[3], Lisa Fogarty[1,2], I-Ting Ho[4] and the SAMI Galaxy Survey Team[5]

[1] Sydney Institute for Astronomy (SIfA), School of Physics, The University of Sydney, NSW 2006, Australia
[2] ARC Centre of Excellence for All-sky Astrophysics (CAASTRO)
[3] Centre for Astrophysics & Supercomputing, Swinburne University of Technology, Hawthorn, Vic 3122, Australia
[4] Institute for Astronomy, University of Hawaii, 2680 Woodlawn Drive, Honolulu, HI 96822, USA
[5] Full list of team members is available at `http://sami-survey.org/members`
Email: `scroom@physics.usyd.edu.au`

Abstract. The Sydney–AAO Multi-object Integral field spectrograph (SAMI) Galaxy Survey is an ongoing project to obtain spatially resolved spectroscopic observations of ~3400 galaxies by mid-2016. To date, a total of ~1000 galaxies have been observed, making the SAMI Galaxy Survey the largest integral field survey in existence. In July 2014 the early data release for the SAMI galaxy Survey occurred, with over 100 galaxies available to the community. The richness of the SAMI dataset allows a vast array of science. We highlight some of the early science results from the project, including the discovery and analysis of galactic winds, the distribution of fast and slow rotating early type galaxies, and the unification of galaxy scaling relations.

Keywords. galaxies: evolution – galaxies: kinematics and dynamics – galaxies: structure – techniques: imaging spectroscopy

---

1. Introduction

Highly multiplexed spectroscopic galaxy surveys, such as the 2dF Galaxy Survey (Colless *et al.* 2001), the Sloan Digital Sky Survey (SDSS; York *et al.* 2000) and the Galaxy And Mass Assembly (GAMA; Driver *et al.* 2011) survey, are powerful tools for investigating galaxy evolution. However, they are limited by only observing a single aperture in the centre of each galaxy, so are inherently unable to investigate the spatially distributed properties of galaxies. Integral field spectrographs (IFSs) provide this missing information. The spatially resolved spectroscopic data produced by an IFS can be used to measure an enormous range of quantities, including the kinematics of gas and stars, spatially varying ionization states, stellar age and metallicity distributions, the location of current star formation and much more. Until recently, technical limitations made IFS observations of large samples time-consuming, as each object was targeted one at a time. Despite this, spectacular scientific progress has been made with samples such as ATLAS-3D (Cappellari *et al.* 2011) and CALIFA (Sánchez *et al.* 2012).

A major leap forward is now possible due to instruments that can make IFS observations of multiple objects at once. The Sydney–AAO Multi-object Integral field spectrograph (SAMI) is one of the first instruments in this class, with thirteen 15″diameter integral field units (IFUs) deployed across a 1-degree field (Croom *et al.* 2012). Each IFU consists of a bundle of 61 optical fibres lightly fused to have a high (~75%) filling factor (Bryant *et al.* 2014a). SAMI is installed on the 3.9-m Anglo-Australian Telescope, feeding the existing AAOmega spectrograph (Sharp *et al.* 2006).

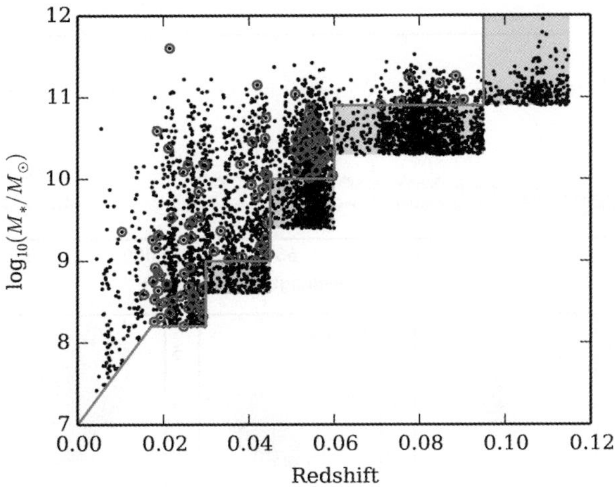

Figure 1. Stellar mass and redshift for all galaxies in the GAMA regions of the SAMI Galaxy Survey (black points) and those in the EDR sample (red circles). The area above the blue boundary indicates the primary sample, while the shaded regions indicate lower-priority targets.

2. The SAMI Galaxy Survey

The SAMI Galaxy Survey will obtain integral field spectroscopic data for ~3400 galaxies, with an expected completion date of mid-2016. At the time of writing, ~1000 galaxies have been observed, including the pilot survey of ~100 galaxies. The overarching science questions for the project are:

• What are the primary physical mechanisms causing environmental suppression of star formation and morphological changes in galaxies?

• How common are inflows and outflows of gas and how do they influence galaxy properties?

• How is angular momentum distributed in the galaxy population and how is mass built up?

AAOmega is a double-beam spectrograph, and for our project we use a 580V grating in the blue arm, giving a resolution $R \sim 1700$ over 3700–5700Å, while the 1000R grating in the red arm results in $R \sim 4500$ over 6300–7400Å. This setup provides an excellent balance, accessing the many emission and absorption features in the blue, but also providing enhanced resolution around the Hα and [N II] emission lines in the red.

The target galaxies for the SAMI Galaxy Survey are split into two samples: a volume-limited field and group sample from the Galaxy And Mass Assembly (GAMA; Driver *et al.* 2011) survey G09, G12 and G15 fields (see Fig. 1), and a cluster sample from a set of eight galaxy clusters. The inclusion of a dedicated cluster sample extends the range of environmental densities to higher values than a simple mass- or luminosity-selected sample of galaxies would provide. Full details of the target selection are provided by Bryant *et al.* (2014b).

The details of SAMI data reduction is described by Sharp *et al.* (2015). This is based on the AAO's 2DFDR† package, together with a dedicated pipeline which handles flux calibration and combining dithered observations into cubes. Fig. 2 illustrates the finished product for a single galaxy, showing the nature and quality of the data obtained.

We have recently made available a set of fully calibrated datacubes for 107 galaxies, forming the SAMI Galaxy Survey Early Data Release (EDR; Allen *et al.* 2015). The

† `http://www.aao.gov.au/science/software/2dfdr`

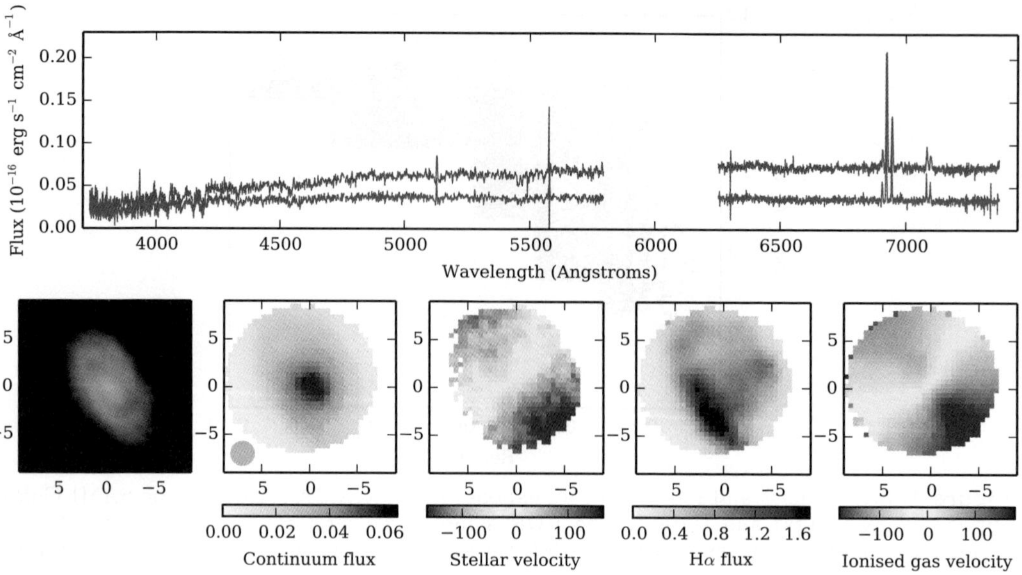

Figure 2. Example SAMI data for the galaxy 511867, with $z = 0.05523$ and $M_* = 10^{10.68} M_\odot$. Upper panel: flux for a central spaxel (blue) and one 4" to the North (red). Lower panels, from left to right: SDSS *gri* image; continuum flux map ($10^{-16}\,\mathrm{erg\,s^{-1}\,cm^{-2}\,\AA^{-1}}$); stellar velocity field ($\mathrm{km\,s^{-1}}$); Hα flux map ($10^{-16}\,\mathrm{erg\,s^{-1}\,cm^{-2}}$); H$\alpha$ velocity field ($\mathrm{km\,s^{-1}}$). The two velocity fields are each scaled individually. Each panel is 18″ square, with North up and East to the left. The grey circle in the second panel shows the FWHM of the PSF.

galaxies were selected from the GAMA regions of the survey. Also available are the datacubes of the corresponding secondary standard stars and a table of ancillary data including quantities such as stellar mass, effective radius and surface brightness. All data can be downloaded from the SAMI Galaxy Survey EDR website‡.

3. First Science

Early science investigations have included studies such as detailed diagnostics of galactic winds (Fogarty *et al.* 2012; Ho *et al.* 2014), study of the kinematic morphology–density relation in clusters (Fogarty *et al.* 2014), star formation in dwarf galaxies (Richards *et al.* 2014) and new analysis of galaxy scaling relations (Cortese *et al.* 2014). Other ongoing work includes studies of spatially resolved star formation and how it is modulated with environment, dynamical mass estimates and the fundamental plane, radial age gradients and much more. Here we highlight just a few of the recent results.

3.1. *Decomposing a star-burst driven wind*

The high spectral resolution of SAMI data in the red arm of AAOmega makes it an excellent probe of complex kinematic structure. Recently Ho *et al.* (2014) demonstrated this, using SAMI to discover and analyse a star-burst driven wind in an isolated disk galaxy. There are multiple kinematics components within this galaxy, which shows (1) a narrow kinematic component consistent with HII regions, (2) a broad kinematic component consistent with shock excitation, and (3) an intermediate component consistent with shock and photoionisation mixing. The three kinematic components have distinct velocity fields, velocity dispersions, line ratios, and electron densities (see Fig. 3). The bipolar galactic winds introduce shock excitation and cause the elevated line ratios and

‡ `http://sami-survey.org/edr`

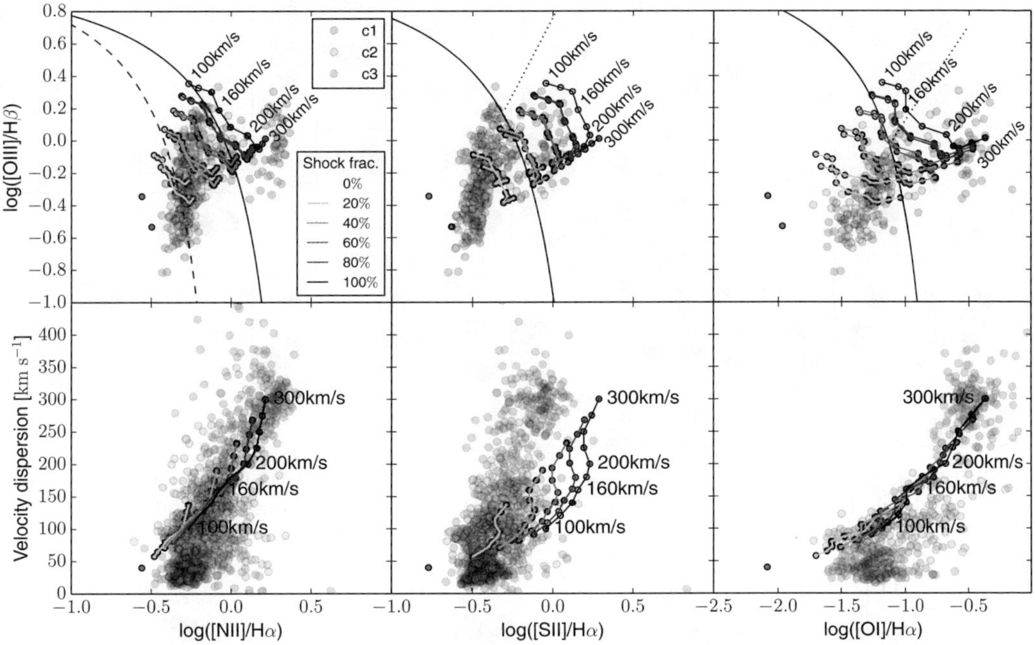

Figure 3. *Top:* The three standard BPT (Baldwin, Phillips & Terlevich 1981) diagrams for a single galaxy from Ho *et al.* (2014) *Bottom:* The three line ratios vs. velocity dispersion. The blue, orange, and red points denote the three distinct kinematic components fit. The points connected by lines are shock/photoionization model grids from `MAPPINGS` (Dopita *et al.* 2013). The lines connect models of a given shock fraction with changing shock velocities. The pure photoionization (zero shock fraction) is at the bottom left.

velocity dispersion in the broad kinematic component. Building on this analysis we are now assessing the frequency of winds and outflows throughout the full SAMI sample.

3.2. *Kinematic morphology in clusters*

Understanding how the distribution of stellar angular momentum is built up is a key focus of the SAMI Galaxy Survey. To do this we are measuring the specific stellar angular momentum (quantified using the λ_r parameter, where $\lambda_r = \langle R|V_{rot}|\rangle / \langle R\sqrt{V_{rot}^2 + \sigma^2}\rangle$; Emsellem *et al.* 2011) and identifying rotation supported galaxies and dispersion supported galaxies. With the large sample size of SAMI we are then able to study how trends in angular momentum vary with galaxy environment, mass and other characteristics. Fogarty *et al.* (2014) find that slow rotators are distributed across a range of environments within clusters, and also find evidence that groups in-falling into clusters contain slow rotators.

3.3. *Galaxy scaling relations*

Galaxy scaling relations such as the Tully-Fisher (1977) relation for spirals, or the Faber-Jackson (1976) relation for early types provide fundamental insights into galaxy structure. Until now, such relations have always been formed from a subset of galaxies, selected on morphology, inclination and other properties. We have taken advantage of the data from the SAMI Galaxy Survey to investigate the relation between the kinematics of gas and stars, and stellar mass in a comprehensive sample of nearby galaxies, regardless of morphology (Cortese *et al.* 2014). We find that regardless of their morphology, all galaxies lie on a tight relation linking stellar mass (M_*) to internal velocity quantified

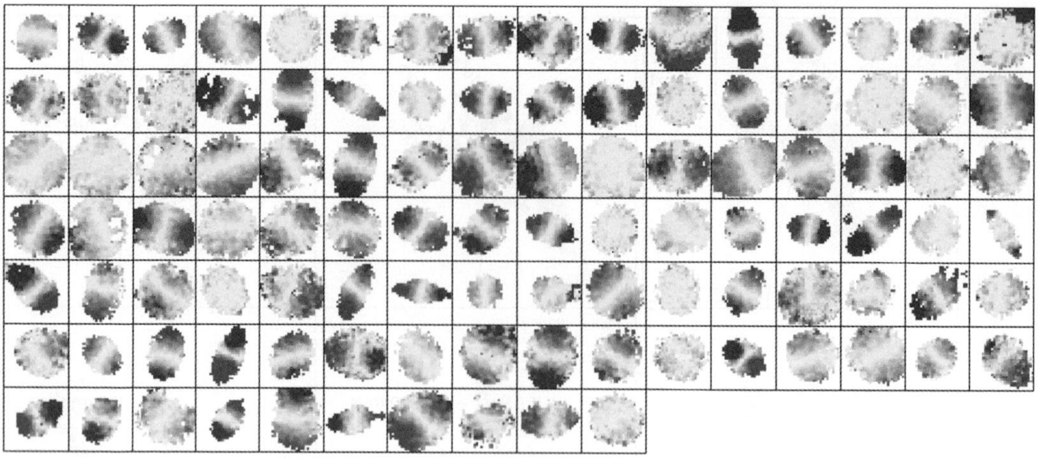

Figure 4. Stellar velocity maps for the 106 galaxies from the SAMI Pilot Survey in three galaxy clusters used by Fogarty *et al.* (2014). The velocity range is $\pm 150 \mathrm{km\,s^{-1}}$

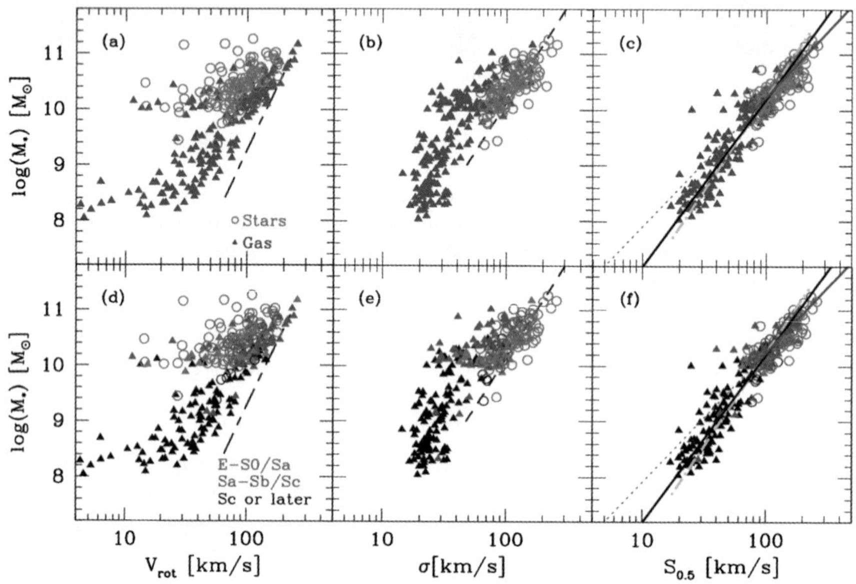

Figure 5. The M vs. V_{rot} (left), σ (center) and $S_{0.5}$ (right) relations for SAMI galaxies from Cortese *et al.* (2014). Circles and triangles indicate stellar and gas kinematics, respectively. In the bottom row, symbols are colour-coded according to morphological type: E-S0/Sa (magenta), Sa-Sb/Sc (dark green), Sc or later types (black). The stellar mass Tully-Fisher (long-short dashed line; Bell & de Jong 2001) and Faber-Jackson (dashed line; Gallazzi *et al.* 2006) relations for nearby galaxies are shown in the left and central panel for comparison. In the right panels, the best inverse linear fit for the whole sample (black line), and for $M > 10^{10}$ M$_\odot$ (brown line) are shown. The dashed-dotted line shows the M vs. $S_{0:5}$ relation obtained by Kassin *et al.* (2007).

by the $S_{0.5}$ parameter (Kassin *et al.* 2007), which combines both the dispersion (σ) and rotational velocity (V_{rot}) contributions to the dynamical support of a galaxy, such that $S_{0.5} = \sqrt{0.5V_{rot}^2 + \sigma^2}$ (see Fig. 5). Galaxies lie on the same relation regardless of whether we use kinematics measured from stars or gas. This relation appears to be more general and at least as tight as any other dynamical scaling relation, representing a unique tool for investigating the link between galaxy kinematics and baryonic content, and a less biased comparison with theoretical models.

4. Conclusions

By completion in 2016 the SAMI Galaxy Survey will have collected integral field data for ~ 3400 galaxies in the local Universe, enabling a vast array of science. We are truly entering an era when integral field observations will be central to our understanding of galaxies. Major new projects are now starting, including MaNGA (see Bundy *et al.* in this proceedings) to extend sample sizes to $\sim 10,000$ in the local Universe. At high redshift KMOS (see Davies *et al.*, in this proceedings) will drive our understanding of galaxy evolution with thousands of galaxies at $z \simeq 1-2$. Next generation facilities, such as Hector (Bland-Hawthorn 2014), will push our understanding in new directions by building samples of $\sim 100,000$ galaxies.

Acknowledgements

The SAMI Galaxy Survey is based on observation made at the Anglo-Australian Telescope. The Sydney-AAO Multi-object Integral field spectrograph was developed jointly by the University of Sydney and the Australian Astronomical Observatory. The SAMI input catalogue is based on data taken from the Sloan Digital Sky Survey, the GAMA Survey and the VST ATLAS Survey. The SAMI Galaxy Survey is funded by the Australian Research Council Centre of Excellence for All-sky Astrophysics, through project number CE110001020, and other participating institutions. The SAMI Galaxy Survey website is `http://sami-survey.org/`.

References

Allen, J. T., *et al.* 2015, *MNRAS*, 446, 1567
Baldwin, J. A., Phillips, M. M., & Terlevich, R., 1981, *PASP*, 93, 5
Bell, E. & de Jong, R., 2001, *ApJ*, 550, 212
Bland-Hawthorn, J., 2014, in B. L. Ziegler, F. Combes, H. Dannerbauer, M. Verdugo, eds, Proc. IAU Symp. 309, Galaxies in 3D across the Universe. Cambridge Univ. Press, Cambridge
Bryant, J. J., Bland-Hawthorn, J., Fogarty, L. M. R., Lawrence, J. S., & Croom, S. M., 2014a, *MNRAS*, 438, 869
Bryant, J. J., *et al.*, 2014b, *MNRAS* submitted, arXiv:1407.7335
Cappellari, M., *et al.*, 2011, *MNRAS*, 416, 1680
Colless, M. M., *et al.*, 2001, *MNRAS*, 328, 1039
Cortese, L., *et al.* 2014, *ApJ*, 795, LL37
Croom, S. M., *et al.*, 2012, *MNRAS*, 421, 872
Dopita, M. A., *et al.*, 2013, *ApJS*, 208, 10
Driver, S. P., *et al.*, 2011, *MNRAS*, 413, 971
Emsellem, E., *et al.*, 2007, *MNRAS*, 379, 401
Faber, S. M. & Jackson R. E., 1976, *ApJ*, 204, 668
Fogarty, L. M. R., *et al.*, 2012, *ApJ*, 761, 169
Fogarty, L. M. R., *et al.*, 2014, *MNRAS*, 443, 485
Gallazzi, A., Charlot, S., Brinchmann, J., & White, S. D. M., 2006, *MNRAS*, 370, 1106
Ho, I.-T., *et al.*, 2014, *MNRAS*, 444, 3894
Kassin, S., *et al.*, 2007, *ApJ*, 660, L35
Richards, S., *et al.*, 2014, *MNRAS*, 445, 1104
Sánchez, S. F., *et al.*, 2012, *A&A*, 538, A8
Sharp, R., *et al.*, 2006, in *SPIE Conference Series* Vol. 6269
Sharp, R., *et al.* 2015, *MNRAS*, 446, 1551
Tully, R. B. & Fisher J. R., 1977, *A&A*, 54, 661
York, D. G., *et al.*, 2000, *AJ*, 120, 1579

Galaxy Masses as Constraints of Formation Models
Proceedings IAU Symposium No. 311, 2014
M. Cappellari & S. Courteau, eds.

doi:10.1017/S174392131500349X

The KMOS Galaxy Clusters Project

Roger L. Davies[1], A. Beifiori[2], R. Bender[2], M. Cappellari[1], J. Chan[2], R. Houghton[1], T. Mendel[2], R.Saglia[2], R. Sharples[3], J. Stott[3], R. Smith[3] and D. Wilman[2]

[1]Sub-department of Astrophysics, Department of Physics, University of Oxford, Denys Wilkinson Building, Keble Road, Oxford OX1 3RH. email: `rld@astro.ox.ac.uk`
[2] Max Planck Institut für Extraterrestrische Physik & University Observatory of the Ludwig-Maximilians-University, Munich.
[3] Department of Physics & Astronomy, Durham University.

Abstract. KMOS is a cryogenic infrared spectrograph fed by twentyfour deployable integral field units that patrol a 7.2 arcminute diameter field of view at the Nasmyth focus of the ESO VLT. It is well suited to the study of galaxy clusters at 1< z < 2 where the well understood features in the restframe V-band are shifted into the KMOS spectral bands. Coupled with HST imagining, KMOS offers a window on the critical epoch for galaxy evolution, 7-10 Gyrs ago, when the key properties of cluster galaxies were established. We aim to investigate the size, mass, morphology and star formation history of galaxies in the clusters. Here we describe the instrument, discuss the status of the observations and report some preliminary results.

Keywords. galaxies: elliptical and lenticular, galaxies: formation, galaxies: clusters

1. Scientific motivation

'Red and dead' quiescent early type galaxies are abundant in local rich clusters of galaxies where star forming, spiral galaxies are rare (Dressler 1980). At z=0.5 the fraction of star forming disk galaxies is higher, (Dressler *et al.* 1997). These results are intrepreted in terms of the transformation of spiral to S0 galaxies over the 4.5Gyr interval. We thus expect the diversity of the Hubble sequence to be established at 0.5<z<2 during the peak epoch of star formation (Madau *et al.* 1996, 1998). Studying high density environments at these redshifts is imperative to reveal the mechanisms that quench star formation and generate the distribution of morphological types we see in local clusters today. Mechanisms that have been suggested for quenching star formation include ram pressure stripping, (Gunn & Gott 1972), harassment, (Moore *et al.* 1996), and strangulation (Larson *et al.* 1980).

The evolution of early-type galaxies can be studied through measuring their scaling relations, such as the Fundamental Plane, as a function of redshift. Several studies have revealed that massive ETGs are in place at z$\sim$2 (e.g. Cimatti *et al.* 2004) but with smaller sizes (e.g. Trujillo *et al.* 2007; Newman *et al.* 2012) and higher stellar velocity dispersions (e.g. Cappellari *et al.* 2009; Saglia *et al.* 2010; Toft *et al.* 2012; Onodera *et al.* 2012; van de Sande *et al.* 2013) compared to local counterparts. However, these authors disagree on the degree of evolution in size (Mancini *et al.* 2010; Szomoru *et al.* 2011) and velocity dispersion (van Dokkum *et al.* 2009; Cappellari *et al.* 2009; Onodera *et al.* 2012), which prevents us from identifying the true cause of these changes.

Furthermore locally we observe that the stars in the most massive galaxies formed at the highest redshifts (Thomas *et al.* 2005) and, according to [α/Fe] excess, over short timescales from high metallicity gas. This shift of active star formation from high mass to low mass galaxies is known as 'downsizing' (Cowie *et al.* 1996). Although the dynamical

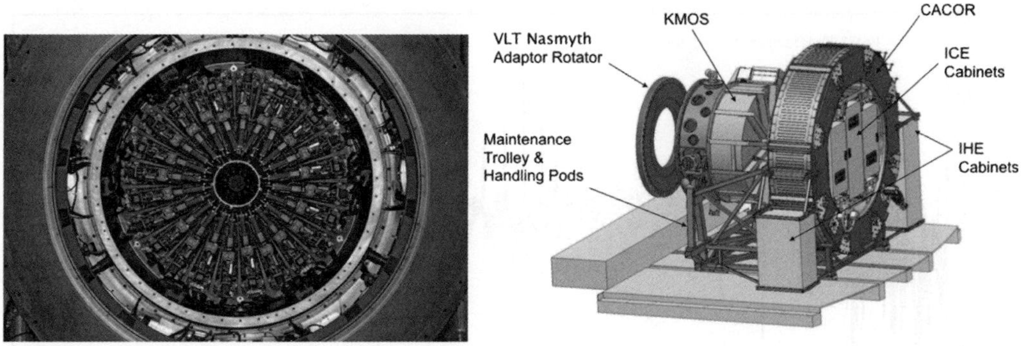

Figure 1. The left hand image shows the 24 pickoff arms of KMOS in their retracted positions. The right hand schematic illustrates the position of KMOS on the Nasmyth platform of the VLT.

M/L evolution of the Fundamental Plane up to z∼1 is consistent with passive evolution (Barr *et al.* 2006; Jørgensen *et al.* 2006; van Dokkum & van der Marel 2007; Saglia *et al.* 2010), the presence of downsizing in the FP is strongly contested (Fritz *et al.* 2009; Holden *et al.* 2010).

It has been proposed that the evolution of ETGs is driven by a combination of their merging history and mass-to-light evolution (Toft *et al.* 2012). Many mechanisms have been proposed such as galaxy growth by major/minor and/or dry/wet mergers and even secular-like "puffing" via AGN feedback (e.g. Hopkins *et al.* 2009, 2010; Oser *et al.* 2012; Hilz *et al.* 2012). Understanding which of these mechanisms drives the evolution of the scaling relations to low-z relies on measuring sizes and velocity dispersions for a 'fair' sample of galaxies at high-z. To date this has not proved possible because, although deep, high-quality imaging data are available for numerous z>1 clusters, current z∼1 spectroscopic studies have sampled only the brightest and most massive galaxies in a few clusters, rarely extending below $10^{11} M_{\odot}$ and typically selecting only a few galaxies per cluster (van Dokkum & van der Marel 2007)). Dynamical studies at z>1.3 only exist for few tens of the brightest objects (van Dokkum *et al.* 2009; Toft *et al.* 2012; van de Sande *et al.* 2013; Belli *et al.* 2014), or for stacked spectra (Cappellari *et al.* 2009).

With the advent of KMOS many of these impediments to the study of high redshift clusters have been removed. The factor of 24 multiplexing allows more than an order-of-magnitude increase in observing efficiency, the IFU eliminates slit losses and the latest NIR detectors are as efficient as CCDs beyond 0.8μm, allowing us to observe at z>1. We are investing significant GTO resources to undertake a deep spectroscopic survey of large numbers of galaxies ($>$ 80), which heretofore would not be feasible. Using this sample we aim to (1) exploit the FP and dynamical M/L studies at z∼1.5 to constrain formation ages; (2) measure the evolution in size and velocity dispersion of a large, representative sample of individual galaxies, to robustly distinguish between different merger history models, (3) use well-understood models of the optical indices to measure stellar ages and metallicities at a time where differences are pronounced and free from low-z residual SF (Kaviraj *et al.* 2007, 2008) and (4) characterise the evolution of age, metallicity and size along the FP (Graves *et al.* 2009; Graves & Faber 2010; Saglia *et al.* 2010; Toft *et al.* 2012; Bezanson *et al.* 2013; van de Sande *et al.* 2014) to reveal how ETGs are assembled in the cluster environment.

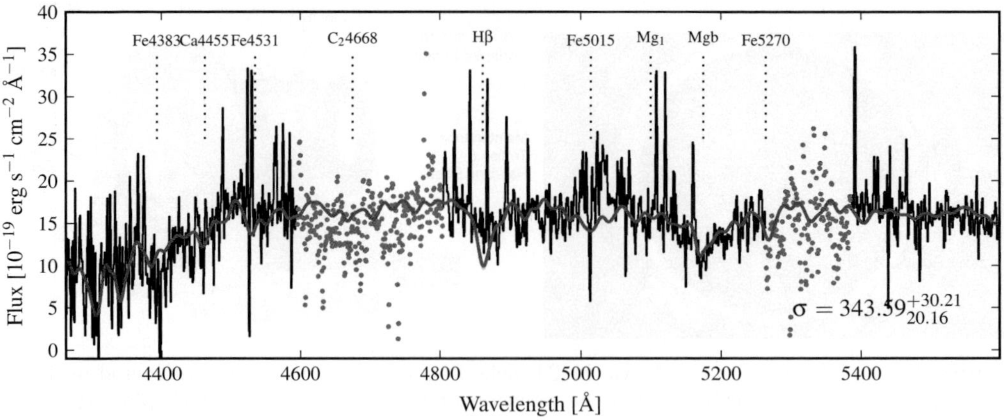

Figure 2. A spectrum of the H=21 galaxy 576 from the z=1.39 cluster XMMU J2235-2557. The on-source exposure time is 11 hours. The spectrum is extracted in a 1.5 arcsecond diameter aperture. The red line superimposed on the spectrum is a fit, using the MILES stellar library (Sánchez-Blázquez *et al.* 2006), indicating a velocity dispersion of 344kms^{-1}.

2. The KMOS instrument

KMOS is an infrared instrument comprising three spectrometers in a single cryostat each fed from eight image sliced integral field units. Each IFU has 14 × 14 spaxels each spanning 2.8 arcseconds square on the sky. They can be deployed within a 7.2 arcminute field of view at the Nasmyth focus of the VLT. The instrument therefore has the potential to observe 24 galaxies simultaneously with an array of 196 spaxels on each galaxy. The 24 retracted arms are shown in the focal plane in Figure 1 where the layout of the instrument on the Nasmyth platform is also shown. The spectrometers provide wavelength coverage from 0.8-2.5μm in each of IZ, YJ, H & K at a resolution of 3300-3800 to facilitate accurate sky subtraction in the wavelength ranges dominated by strong sky emission lines. The efficiency of the spectrograph is $>$ 20% in YJ and $>$ 30% in H and K. More details of the technical performance can be found in (Sharples *et al.* 2014) and some early commissioning and science verification results are presented in (Sharples *et al.* 2013). The instrument was commissioned during 2013 and we started our cluster observations, which use the YJ bands, in October 2013.

3. The selection of clusters and galaxies

We selected three clusters for study: XMMU J2235-2257 (z=1.39), XMMXCSJ2215.9-1738 (z=1.46) and Cl0332-2742 (z=1.61), each of which has a significant amount of archive data available and a rich enough ETG population to address our science goals. As a backup for when the wind is from the north, or in less optimal observing conditions we selected RCS234526-3632.6 (z=1.04) which has the CaH+K, Hγ, and V-band indices in the KMOS IZ band. Multiband HST photometry is available for all clusters including existing or upcoming WFC3 IR imaging from our HST Cycle 22 program. This is crucial for measuring the sizes of galaxies, mapping the distribution of stellar mass and measuring their structural parameters (which are necessary to construct the Fundamental Plane in stellar mass and light), and determining morphologies. Moreover, deep ground based imaging allows us to identify samples of galaxies on the red sequence and fit spectral energy distributions. Galaxies that have spectroscopically confirmed cluster membership increases our selection efficiency. At z$\sim$1.4 & z=1.6 the key absorption features (Balmer,

Mg & Fe lines) are relatively uncontaminated by strong sky emission or telluric absorption. The KMOS patrol field is well matched to the extent of the clusters on the sky and the IFUs match the sizes of the galaxies. Our cluster sample spans a range of cluster mass, temperature, and morphology, so we can study the clusters through comparison with simulations.

Individual galaxies were initially selected from pre-existing (spectroscopic) membership catalogs, and then from the red sequence below J(AB)$<$ 22.5 mag or the equivalent in H-band. We fill unallocated IFUs with blue star-forming objects: confirmed cluster members first, then above the mass limit imposed by the red sequence magnitude limit.

4. Early results

Figure 2 shows the spectrum of one of the brighest galaxies, #576, in XMMUJ2235-2557 summed over an aperture of diameter 1.5 arcseconds obtained after 11 hours of on-source integration. Absorption features of associated with the MgH bandhead at 5200$\AA$ are apparent as is Hβ. The model spectrum overplotted is from the MILES library broadened to a velocity dispersion of 344kms^{-1}. Although it is possible to derive velocities and dispersions for the brightest galaxies in the cluster it will be necessary to stack spectra for the faintest galaxies. As yet we do not have sufficient high quality spectra to explore the Fundamental Plane.

The power of the IFU to reveal physical structure that is hard to extract from a slit spectrum is illustrated in Figure 3, which shows, on the left, HST WFC3 F110W images of three galaxies in XMMUJ2235-2557 and on the right the stellar continuum, shown in red, and in two cases emission from the ionised gas ([OIII] 5007$\AA$), shown in green. Between them is the ESO/VLT HAWK-I J & K composite image of the cluster showing the location of the three galaxies. All three are some way from the centre of the cluster and outside the region of X-ray emission shown in (Mullis *et al.* 2005). For the two galaxies showing emission, the ionised gas is offset significantly from the stellar continuum indicating that this ionised gas is either flowing out of the galaxy or swept back by the motion of the galaxy through the intracluster medium, even at the low densities implied by the X-ray emission. We are investigating these possibilities further.

5. Conclusions

The multi-IFU infrared spectrometer, KMOS, has been brought into operation on the ESO VLT and we are using it to study the evolution of galaxies in rich clusters with 1$<$ z $<$ 2. At these redshifts the well understood V-band absorption features are shifted into the infrared and are detected in the individual spectra of the brightest galaxies. Coupled with the HST imaging available for the clusters we will study the evolution of the ETG scaling relations and through that the mass-to-light ration evolution. The multiple integral field units enable us to examine the spatial distribution of emission at different wavelengths and we have shown that the emission of [OIII] in two galaxies is significantly offset from the centre of the galaxy potential, even in galaxies that the X-ray maps indicate are in regions of low intracluster medium density.

Acknowledgements

This work was supported by the Astrophysics at Oxford grants (ST/H002456/1 and ST/K00106X/1) as well as visitors grant (ST/H504862/1) from the UK Science and Technology Facilities Council. RLD acknowledges travel and computer grants from Christ

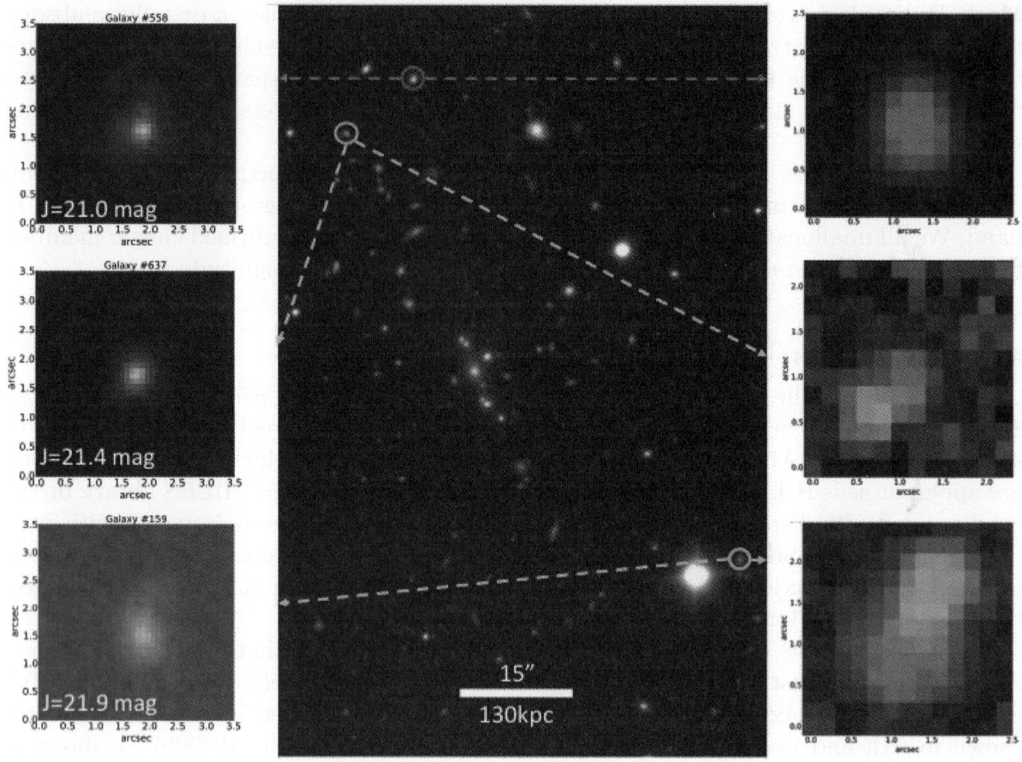

Figure 3. The central panel shows the ESO/VLT HAWK-I J & K composite image of the z= 1.39 cluster XMMUJ2235-2557. On the left are HST WFC3 F110W thumbnail images of three galaxies with J = 21.0 to 21.9. To the right are the same galaxies but shown in two band passes, (i) in the J-band stellar continuum shown in red and (ii) the emission from [OIII] 5007Å shown in green. The lower two galaxies exhibit emission significantly offset from the stellar continuum indicating that the ionised gas has been disturbed. Note that the 'stretch' for the emission line (green) images on the right has been chosen to emphasise the presence of emission.

Church, Oxford. The KMOS project has been supported by the Verbundforschung of the German Ministry for Education and Reseach (BMBF), project number 05A11WMA."

References

Barr, J., Jørgensen, I., Chiboucas, K., Davies, R., & Bergmann, M. 2006, *ApJ*, 649, L1
Belli, S., Newman, A. B., & Ellis, R. S. 2014, *ApJ*, 783, 117
Bezanson, R., van Dokkum, P. G., van de Sande, J., *et al.* 2013, *ApJ*, 779, L21
Cappellari, M., di Serego Alighieri, S., Cimatti, A., *et al.* 2009, *ApJ*, 704, L34
Cimatti, A., Daddi, E., Renzini, A., *et al.* 2004, *Nature*, 430, 184
Cowie, L. L., Songaila, A., Hu, E. M., & Cohen, J. G. 1996, *AJ*, 112, 839
Dressler, A. 1980, *ApJ*, 236, 351
Dressler, A., Oemler, A., Jr., Couch, W. J., *et al.* 1997, *ApJ*, 490, 577
Fritz, A., Böhm, A., & Ziegler, B. L. 2009, *MNRAS*, 393, 1467
Graves, G. J., Faber, S. M., & Schiavon, R. P. 2009, *ApJ*, 698, 1590
Graves, G. J. & Faber, S. M. 2010, *ApJ*, 717, 803
Gunn, J. E. & Gott, J. R., III 1972, *ApJ*, 176, 1
Hilz, M., Naab, T., Ostriker, J. P., *et al.* 2012, *MNRAS*, 425, 3119
Holden, B. P., van der Wel, A., Kelson, D. D., Franx, M., & Illingworth, G. D. 2010, *ApJ*, 724, 714

Hopkins, P. F., Somerville, R. S., Cox, T. J., *et al.* 2009, *MNRAS*, 397, 802
Hopkins, P. F., Bundy, K., Croton, D., *et al.* 2010, *ApJ*, 715, 202
Jørgensen, I., Chiboucas, K., Flint, K., *et al.* 2006, *ApJ*, 639, L9
Kaviraj, S., Rey, S.-C., Rich, R. M., Yoon, S.-J., & Yi, S. K. 2007, *MNRAS*, 381, L74
Kaviraj, S., Khochfar, S., Schawinski, K., *et al.* 2008, *MNRAS*, 388, 67
Larson, R. B., Tinsley, B. M., & Caldwell, C. N. 1980, *ApJ*, 237, 692
Madau, P., Ferguson, H. C., Dickinson, M. E., *et al.* 1996, *MNRAS*, 283, 1388
Madau, P., Pozzetti, L., & Dickinson, M. 1998, *ApJ*, 498, 106
Mancini, C., Daddi, E., Renzini, A., *et al.* 2010, *MNRAS*, 401, 933
Moore, B., Katz, N., Lake, G., Dressler, A., & Oemler, A. 1996, *Nature*, 379, 613
Mullis, C. R., Rosati, P., Lamer, G., *et al.* 2005, *ApJ*, 623, L85
Newman, A. B., Ellis, R. S., Bundy, K., & Treu, T. 2012, *ApJ*, 746, 162
Onodera, M., Renzini, A., Carollo, M., *et al.* 2012, *ApJ*, 755, 26
Oser, L., Naab, T., Ostriker, J. P., & Johansson, P. H. 2012, *ApJ*, 744, 63
Saglia, R. P., Sánchez-Blázquez, P., Bender, R., *et al.* 2010, *A&A*, 524, A6
Sánchez-Blázquez, P., Peletier, R. F., Jiménez-Vicente, J., *et al.* 2006, *MNRAS*, 371, 703
Sharples, R., Bender, R., Agudo Berbel, A., *et al.* 2013, The Messenger, 151, 21
Sharples, R., Bender, R., Agudo Berbel, A., *et al.* 2014, Proc. SPIE, 9147, 91470W
Szomoru, D., Franx, M., Bouwens, R. J., *et al.* 2011, *ApJ*, 735, L22
Thomas, D., Maraston, C., Bender, R., & Mendes de Oliveira, C. 2005, *ApJ*, 621, 673
Toft, S., Gallazzi, A., Zirm, A., *et al.* 2012, *ApJ*, 754, 3
Trujillo, I., Conselice, C. J., Bundy, K., *et al.* 2007, *MNRAS*, 382, 109
van de Sande, J., Kriek, M., Franx, M., *et al.* 2013, *ApJ*, 771, 85
van de Sande, J., Kriek, M., Franx, M., Bezanson, R., & van Dokkum, P. G. 2014, *ApJ*, 793, L31
van Dokkum, P. G. & van der Marel, R. P. 2007, *ApJ*, 655, 30
van Dokkum, P. G., Kriek, M., & Franx, M. 2009, *Nature*, 460, 717

Galaxy Masses as Constraints of Formation Models
Proceedings IAU Symposium No. 311, 2014
M. Cappellari & S. Courteau, eds.

doi:10.1017/S1743921315003506

The Dark Halo – Spheroid Conspiracy Reloaded: Evolution with Redshift

Rhea-Silvia Remus[1], Klaus Dolag[1,2] and Andreas Burkert[1,3]

[1]Universitäts-Sternwarte München, Scheinerstr. 1, D-81679 München, Germany
[2]MPI for Astrophysics, Karl-Schwarzschild Strasse 1, D-85748 Garching, Germany
[3]MPI for Extraterrestrial Physics, Giessenbachstrasse 1, D-85748 Garching, Germany
email: rhea@usm.lmu.de

Abstract. The total density profiles of elliptical galaxies can be fit by a single power law, i.e., $\rho_{\rm tot} \propto r^{\gamma}$ with $\gamma \approx -2$. While strong lensing observations show a tendency for the slopes to become flatter with increasing redshift, simulations indicate an opposite trend. To understand this discrepancy, we study a set of simulated spheroids formed within the cosmological framework. From our simulations we find that the steepness of the total density slope correlates with the compactness of the stellar component within the half-mass radius, and that spheroidal galaxies tend to be more compact at high redshifts than their present-day counterparts. While both these results are in agreement with observations, the observed trend of the total density slope with redshift remains in contradiction to the results from simulations.

Keywords. galaxies: elliptical – galaxies: evolution – galaxies: formation

1. Introduction

At low redshifts, elliptical galaxies are known to consist mostly of an (old) spheroidal stellar component embedded in a dark matter halo. The transition zone, where the dark matter starts to dominate over the stellar component, provides information about the galaxies' formation history; observationally, however, the kinematics in this area are difficult to trace. For spiral galaxies it is long known from observations that the rotation curves are flat out to large radii (Einasto *et al.* 1974, Faber & Gallagher 1979), which is called the dark-halo–disk conspiracy, and detailed dynamical modeling (e.g., Gerhard *et al.* 2001) has demonstrated more recently that between the dark halo and the spheroid of massive, slow rotating elliptical galaxies a similar conspiracy might exist. This conspiracy for spheroidal galaxies has been studied from simulations in detail by Remus *et al.* (2013) (hereafter R13) for a sample of ellipticals from binary mergers and cosmological zoom-in simulations. They find that the total density and velocity dispersion profiles of spheroidal galaxies can be described by single power laws, with slopes of $\gamma \approx -2$ for the density and $\gamma \approx 0$ for the velocity dispersion. They also report a strong trend towards steeper slopes with redshift.

One important observational method for studying the total density profiles of elliptical galaxies is strong lensing: For low redshift ellipticals, the SLACS survey finds that the total density profiles can be described by a power law with a slope of $\gamma \approx -2$ (Auger *et al.* 2010, Barnabè *et al.* 2011), with the steepness of the slope correlating with the central density of the stellar component and the effective radius, which agrees well with the result from R13. However, for higher redshifts up to $z \approx 1$, strong lensing studies report a slight trend towards flatter slopes (Treu & Koopmans 2004, Ruff *et al.* 2011, Bolton *et al.* 2012, Sonnenfeld *et al.* 2013) in contradiction to the trends found by R13. We present a brief preview of a more detailed study on the evolution of the dark-halo–spheroid conspiracy with redshift and its implications for the formation of elliptical galaxies (Remus *et al.*, in

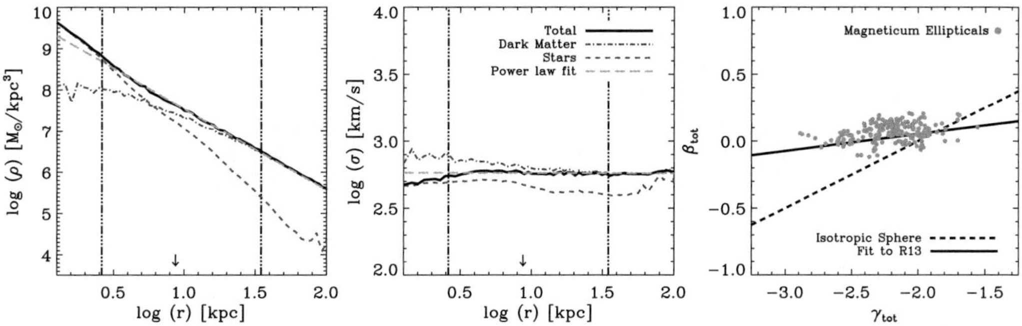

Figure 1. *Left:* Density profiles for a spheroidal from Magneticum: total combined stellar and dark matter profile (black solid line), stellar component (red dashed line), dark matter component (blue dash-dotted line), and a power-law fit to the total density profile (green long-dashed line). The arrow marks $R_{1/2}$. *Center:* Velocity dispersion profile for the same spheroidal. Colors as in the left panel. *Right:* Slopes of the total velocity dispersion profiles $\beta_{\rm tot}$ vs. those of the total density profiles $\gamma_{\rm tot}$ for all spheroidals selected from the simulation. Dashed black line: solution of the Jeans Equation for an isotropic sphere in hydrostatic equilibrium. Solid black line: fit to the zoom-in simulations from R13.

prep.), using spheroidals selected from a full hydrodynamic cosmological box with high resolution called Magneticum Pathfinder (Dolag *et al.*, in prep.).

2. The Magneticum Pathfinder Simulations

The Magneticum Pathfinder simulations* (Dolag *et al.*, in prep.) are a set of hydrodynamical cosmological boxes with different resolutions ($m_{\rm Gas} = 2.6 \cdot 10^9 M_\odot/h$ up to $m_{\rm Gas} = 3.9 \cdot 10^5 M_\odot/h$) and volumes covering $(896\ {\rm Mpc}/h)^3$ to $(18\ {\rm Mpc}/h)^3$, performed with a modified version of GADGET (Springel 2005). They include state-of-the-art physics, especially black hole feedback, and improvements for smooth particle hydrodynamics (for detailed descriptions see Hirschmann *et al.* 2014, Dolag *et al.* 2005, Dolag *et al.*, in prep., and Beck *et al.*, in prep.). We adopt a ΛCDM cosmology according to WMAP7 (Komatsu *et al.* 2011) with $h = 0.704$, $\Omega_m = 0.272$ and $\Lambda = 0.728$, and use SUBFIND (Springel *et al.* 2001, Dolag *et al.* 2009) to identify and extract structures.

We use the largest Magneticum box with enough resolution to study the dynamics of spheroidal galaxies in detail. It has a box length of 48 Mpc/h and a resolution of $m_{\rm DM} = 3.6 \cdot 10^7 M_\odot/h$, $m_{\rm Gas} \approx 7.3 \cdot 10^6 M_\odot/h$, and slightly smaller stellar mass, with $(2 \cdot 576)^3$ particles and a softening length of $\epsilon_{\rm DM,Gas} = 1.4$ kpc/h for dark matter and gas and $\epsilon_* = 0.7$ kpc/h for stars. This simulation successfully reproduces populations of disk as well as spheroidal galaxies. To ensure sufficient resolution, we exclude galaxies with stellar masses less than $M_* = 1 \cdot 10^{11} M_\odot$. We use a procedure similar to Scannapieco *et al.* (2008) to select spheroidals without cold gas disks from the sample of galaxies (see Remus *et al.* (2014) for more details). A more advanced study including spheroidals with small gas disks and S0-galaxies is in preparation (Remus *et al.*).

3. The Dark Halo – Spheroid Conspiracy with Redshift

The total density and velocity dispersion profiles of spheroidal galaxies can be fit by single power laws within a range of $0.3R_{1/2}$ to $4R_{1/2}$, as shown by R13 for a set of cosmological zoom-in simulations and isolated binary mergers. The inner parts of

www.magneticum.org

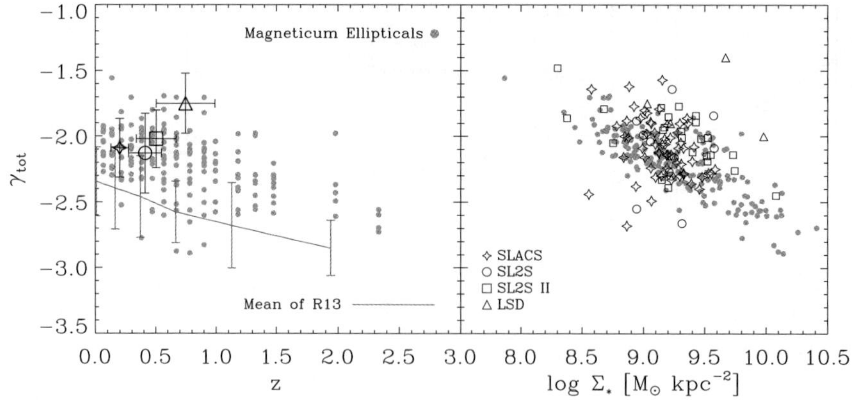

Figure 2. *Left:* Total density slope versus redshift. Filled lilac circles are the Magneticum spheroidals. The red solid line marks the mean of the values from R13, the open symbols the means of the observations from SLACS (diamonds, Auger *et al.* 2010 and Barnabè *et al.* 2011), SL2S (circles from Ruff *et al.* 2011, squares from Sonnenfeld *et al.* 2013) and LSD (triangles, Treu & Koopmans 2004), with the error bars the square roots of the variance. *Right:* Total density slope versus stellar mass density. Symbols and colors as in the left panel, but with the open symbols representing the individual observations from the different surveys.

the total profiles are dominated by the stellar component, while in the outer parts the dark matter contributes most to the profiles. We find the same behaviour for the Magneticum spheroidals, as demonstrated in the left and central panel of Fig. 1 for an example spheroidal. The slopes of the power-law fits,

$$\beta = \frac{\mathrm{d}\log(\sigma)}{\mathrm{d}\log(r)} \quad \text{and} \quad \gamma = \frac{\mathrm{d}\log(\rho)}{\mathrm{d}\log(r)},$$

are shown in the right panel of Fig. 1. We find that the slopes do not follow the solution of the Jeans Equation for an isotropic sphere (or a sphere with constant anisotropy) in hydrostatic equilibrium assuming a single power-law solution for both density and velocity dispersion, but show constant values for β, while γ covers the whole range between -3 and -1.5. This is in agreement with the results from R13, who showed that β is constant independent of the spheroidals' mass, size, in-situ or central dark matter fraction, while the steepness of the total density profiles correlates with the amount of stars formed in-situ and anticorrelates with the dark matter fraction.

However, while R13 found a steepening of γ with redshift, observations appear to indicate a different behaviour. The left panel of Fig. 2 shows γ for the Magneticum spheroidals for different redshifts, together with the observed trends and the results from R13. While the steepening of γ with redshift is not as strong as for the galaxies in R13, we still find the tendency towards steeper slopes for higher redshifts, in contradiction to observations. The difference between the two simulated samples of galaxies is partly due to the improvements to SPH and the additional physics included in the Magneticum simulations, which cause the galaxies to be more realistic (see, for example, the mass–size relation shown by Remus *et al.* 2014), but primarily an effect of the mass selection criterion applied to the Magneticum samples: while R13 followed a set of 20 galaxies back in time, the Magneticum sample includes spheroidals above $M_* = 1 \cdot 10^{11} M_\odot$ independent of their progenitors and thus includes more massive systems at high redshifts than the sample studied by R13. The right panel of Fig. 2 shows the dependence of γ on the stellar mass density $\Sigma_* = M_*/2\pi R_{1/2}^2$, following Sonnenfeld *et al.* (2013). Our Magneticum spheroidals are in very good agreement with the observations, with a clear trend towards steeper slopes for more compact spheroids. Since spheroidals at higher redshifts are generally more compact than their present day counterparts in mass (e.g., van de Sande *et al.* 2013), this supports the idea that the total density slopes γ should be generally steeper with redshift. Therefore, we conclude that it is crucial to understand the origin of the differences between the observed and simulated trends of the total density slope with redshift. A detailed study on this subject will be presented in a forthcoming paper by Remus *et al.* (in prep.).

4. Conclusions

The total combined density profiles of Magneticum spheroidal galaxies can be fit by a single power law with a slope of on average $\gamma \approx -2$, with a tendency towards steeper slopes for more compact ellipticals. At higher redshifts, the density slopes of our spheroidals are generally steeper, in agreement with results presented in R13. While observations agree with the simulations, that spheroidal galaxies tend to be more compact at high redshifts than their present-day counterparts, the observed flattening of the total density slope with redshift contradicts the results from both the simulations presented in R13 and the Magneticum spheroidals presented in this work. The origin of this different behaviour will be discussed in a forthcoming paper by Remus *et al.* (in prep.). In addition, we showed that the slope of the total velocity dispersion is constant independent of the stellar mass density and the total density slopes, in agreement with simulations by R13. This behaviour is still an unsolved problem, and will be subject to further investigations.

Acknowledgements

The authors acknowledge support from the DFG Excellence Cluster Universe, the Computational Center C2PAP, and the Leibniz-Rechenzentrum via projects pr86re and pr83li.

References

Auger, M. W., *et al.* 2010, *ApJ*, 724, 511
Barnabè, M., *et al.* 2011, *MNRAS*, 415, 2215
Bolton, A. S., *et al.* 2012, *ApJ*, 757, 82
Dolag, K., Vazza, F., Brunetti, G., & Tormen, G. 2005, *MNRAS*, 364, 753
Dolag, K., Borgani, S., Murante, G., & Springel, V. 2009, *MNRAS*, 399, 497
Einasto, J., Kaasik, A., & Saar, E. 1974, *Nature*, 250, 309
Faber, S. M. & Gallagher, J. S. 1979, *ARA&A*, 17, 135
Gerhard, O., Kronawitter, A., Saglia, R. P., & Bender, R. 2001, *AJ*, 121, 1936
Hirschmann, M., *et al.* 2014, *MNRAS*, 442, 2304
Komatsu, E., *et al.* 2011, *ApJS*, 192, 18
Remus, R.-S., Burkert, A., & Dolag, K., *et al.* 2013, *ApJ*, 766, 71
Remus, R.-S., *et al.* 2014, in *IAU Symp. 309, Galaxies in 3D across the Universe*, eds. B. L. Ziegler, F. Combes H. Dannerbauer & M. Verdugo, in press
Ruff, A. J., *et al.* 2011, *ApJ*, 727, 96
Scannapieco, C., *et al.* 2008, *MNRAS*, 389, 1137
Sonnenfeld, A., Treu, T., Gavazzi, R., *et al.* 2013, *ApJ*, 777, 98
Springel, V. 2005, *MNRAS*, 364, 1105
Springel, V., White, S. D. M., Tormen, G., & Kauffmann, G. 2001, *MNRAS*, 328, 726
Treu, T. & Koopmans, L. V. E. 2004, *ApJ*, 611, 739
van de Sande, J., Kriek, M., Franx, M., *et al.* 2013, *ApJ*, 771, 85

Galaxy Masses as Constraints of Formation Models
Proceedings IAU Symposium No. 311, 2014
M. Cappellari & S. Courteau, eds.

doi:10.1017/S1743921315003518

Applying galactic archeology to massive galaxies using deep imaging surveys

Pierre-Alain Duc

AIM Paris-Saclay, CEA:Irfu, CNRS/INSU, Université Paris Diderot, France. email: paduc@cea.fr

Abstract. Various programs aimed at exploring the still largely unknown low surface brightness Universe with deep imaging optical surveys have recently started. They open a new window for studies of galaxy evolution, pushing the technique of galactic archeology outside the Local Group (LG). The method, based on the detection and analysis of the diffuse light emitted by collisional debris or extended stellar halos (rather than on stellar counts as done for LG systems), faces however a number of technical difficulties, like the contamination of the images by reflection halos and Galactic cirrus. I review here the on-going efforts to address them and highlight the preliminary promising results obtained with a systematic survey with MegaCam on the CFHT of nearby massive early-type galaxies done as part of the ATLAS^{3D}, NGVS and MATLAS collaborations.

Keywords. galaxies: elliptical and lenticular, cD - galaxies: evolution - galaxies: formation

1. Introduction

The processes responsible for galaxy evolution, their formation, growth and death, may be be studied either probing slices of the Universe at different epochs through redshift surveys or scrutinizing and dissecting todays galaxies to infer their past mass assembly. The success of the latter method strongly depends on our ability to detect structures located in the very outskirts of galaxies: the extended stellar halos and collisional debris that keep the memory of past mergers. Their number and shape inform us on the mechanisms driving the growth of galaxies: wet vs dry, minor vs major mergers, secular processes, etc. Unfortunately, such structures are of low surface brightness (LSB), and are difficult to detect unless their are resolved into individual stars, like for Local Group systems. In the last decades, thanks in particular to the great image quality of the Hubble Space Telescope and dedicated ground-based surveys, galactic archeology made tremendous progress, and the Gaia space mission will boost it even more. However the method is confined to galaxies located within a few Mpc from the Milky Way. The most massive ones, in particular lenticulars and ellipticals the origin of which remains actively debated, are located too far away to be mapped with stellar counts. Until Extremely Large Telescopes become available, the only way to detect their outer structures is through their diffuse light. Fortunately, the availability of large field of view optical cameras, and the development of clever observing and data reduction techniques, have led to a gain of several magnitudes in surface brightness limits with respect to classical imaging and now allow us to detect this diffuse light.

2. On-going deep imaging surveys

2.1. *Achievements: an increased sensitivity to LSB structures*

Thanks to its large coverture of the sky, homogeneity and user friendly access to its archives, the Sloan Digital Sky Survey (SDSS, York *et al.* 2000), has become *the* unavoidable resource for optical studies of galaxies in the nearby Universe. However, its sensitivity to extended low surface brightness features is very limited, largely hampering their direct use for galaxy archeology. A number of on-going projects aim at getting much deeper images. They make use of stacking techniques (e.g. Kim & Im 2013), refurbished or dedicated instruments (e.g. Mihos *et al.* 2005; van Dokkum *et al.* 2014), extremely long exposures on small telescopes (e.g. Martínez-Delgado *et al.* 2010), or LSB-optimized surveys, such as the Next Generation Virgo Cluster Survey (NGVS, Ferrarese *et al.* 2012), or the MegaCam component of the ATLAS^{3D} project (Duc *et al.* 2014), to be completed thanks to the MATLAS CFHT Large Programme.

Fig. 1 illustrates the gain of the LSB-optimized MegaCam survey, which reaches about 28.5–29 mag arcsec^{-2} in the g–band to be compared to about 26.5 for the SDSS.

One should note that the sensitivity to LSB structures does not only depend on the telescope size and exposure time, but is largely driven by the handling of systematics in the observations. Consequently, the limiting surface brightness is particularly difficult to estimate, partly explaining the large range of published values in the literature (see Duc *et al.* 2014, and references therein).

Whether deep ground-based imaging surveys can reach the limiting surface brightness obtained in Local Galaxies with stellar counts, i.e. 32 mag arcsec^{-2} , as claimed for instance for the amazing Dragonfly experiment (Abraham *et al.*, this volume), still needs to be clarified.

2.2. *Challenges: halos and cirrus*

In addition to the classical issues of large-scale variations of the background (not easily) addressed by various flat–fielding or sky subtraction techniques, deep imaging is hampered by two stumbling blocks that considerably limit their capability of detecting diffuse extended structures: artificial halos due to internal reflections in the instrument and Galactic cirrus emission.

The light of bright stars but also the nucleus of galaxies is reflected between the CCD and the different optical elements of the camera, causing the presence on deep images, such as the ones produced by MegaCam, of multiple more or less concentric halos, with sizes reaching several arcmin. As illustrated in Fig. 2, they may hide fine structures like tidal tails or be mistaken with the stellar halos that surround galaxies, or the intra-cluster light in which cluster galaxies are embedded. A special coating of CCDs and filters or a completely different optical construction, as done for the Dragonfly telephoto lenses, may lower the impact of such instrumental signatures. Besides, there are on-going efforts to a posteriori remove them thanks to a modeling of the instrument and ray tracing experiments.

Because it is unavoidable and extremely difficult to subtract, an even more serious issue is the pollution by Galactic cirrus. Dust clouds around our own Milky Way scatter light in the optical regime. They show up on deep images as complex, extended, multicolored, filamentary structures that may be mistaken with stellar tidal tails (see examples in Duc *et al.* 2014, and in Fig. 1). The Planck survey revealed that Galactic cirrus is present over a large fraction of the sky. Cirrus emission may thus potentially be an issue for all deep imaging surveys. In that respect, it is somehow surprising or reassuring that the

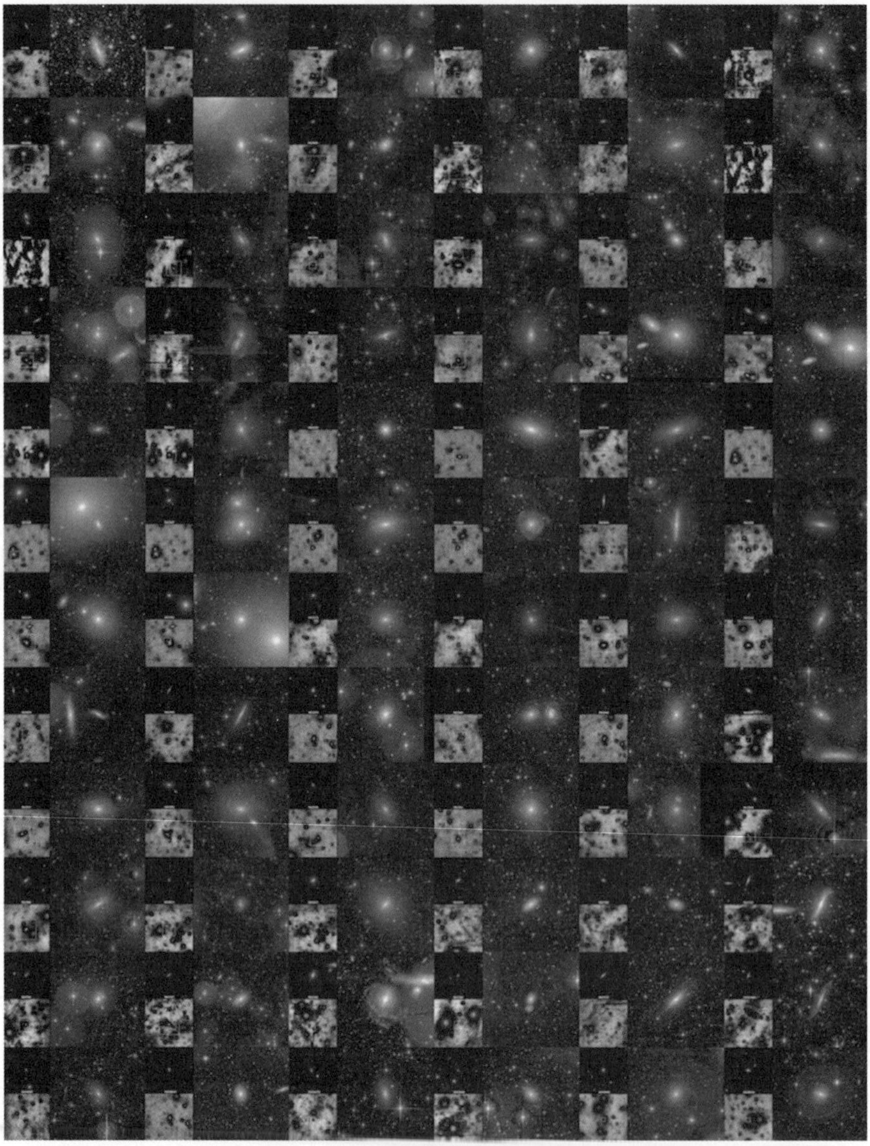

Figure 1. A catalog of early-type galaxies from the ATLAS^{3D} survey. Each sub-panel displays a galaxy as seen by the SDSS (top-right) and the MegaCam/CFHT camera (right). A grey scale g-band MegaCam image of the field around the ETG is shown to the bottom-left. A high resolution version of this poster may be obtained at http://irfu.cea.fr/Projets/matlas/MATLAS/Outreach/Outreach.html

extremely deep image of M101 obtained with Dragonfly (van Dokkum *et al.* 2014) does not show clear evidence of cirrus emission.

3. Results

Despite the technical issues mentioned above, the asset of deep imaging of galaxies is undeniable. It reveals a wealth of so far unknown structures that inform us about the various events that build galaxies, but also raises issues on how galaxies are classified.

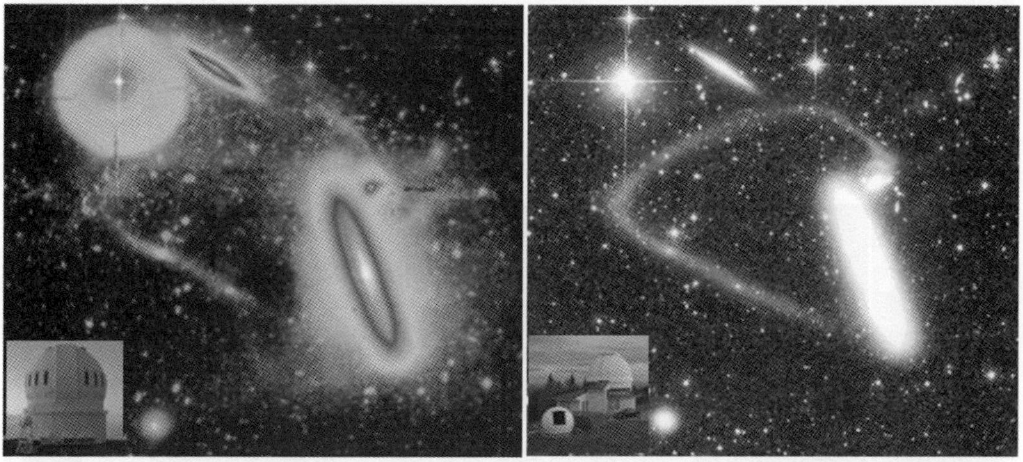

Figure 2. Comparison between the image of the spiral galaxy NGC 4216 obtained with MegaCam at the CFHT as part of the NGVS (Paudel *et al.*, 2013, left, 1 hour exposure time) and at the amateur Irida observatory in Bulgaria (courtesy V. Popov and E. Ivanov, right, 20 hours exposure time). The latter image is not as deep as the first one, but looks cleaner due to the absence of the prominent reflection halos that plague the CFHT image.

3.1. *Is traditional morphological classification challenged by deep imaging?*

The classification of galaxies traditionally relies on morphological criteria (bulge to disk ratio, bar strength) that are checked inspecting optical images. However, the detectability of the discriminating structures depends on the depth of the available images. Indeed, as shown on Fig. 3, some apparently red galaxies classified as early-type by a visual inspection of SDSS images exhibit on the deeper MegaCam images so-far undetected low surface brightness blue spiral-like structures. This does not imply that morphological classification is unfounded, but obliges us to take into consideration on which data set it was done.

3.2. *Are scaling relations affected by deep imaging?*

Stellar mass and sizes are other fundamental parameters used in categorizing galaxies and/or determining universal scaling relations that blur their differences (see example of the mass–size relation of ETGs in Fig. 4). The values of such parameters depend again in principle on the depth of the images used to determine them. We quantified this measuring the diffuse light component beyond an isophote of 26 mag arcsec^{-2} , i.e. the component not detected in traditional shallow imaging surveys. The extra light ranges between 5 and 20 percent (in the g–band), depending on the galaxy mass. It is higher in dense environments such as the Virgo cluster, but there the light of the very extended outer halo mixes with the intra-cluster light, and its importance is difficult to estimate.

For galaxies with masses below 10^{11} $M_{\odot}$, we do not see significant differences between the effective radius of the ATLAS^{3D} ETGs determined with MegaCam and that published the literature. However a significant systematic excess of R_e of up to a factor of 1.7 is found above this threshold (Duc *et al.* 2014).

Such updates of the mass and size do not significantly alter the global scaling relations given their intrinsic scatter. However, the variations of the mass/size excess as a function of mass, morphological type, etc., may give clues on how galaxies were assembled.

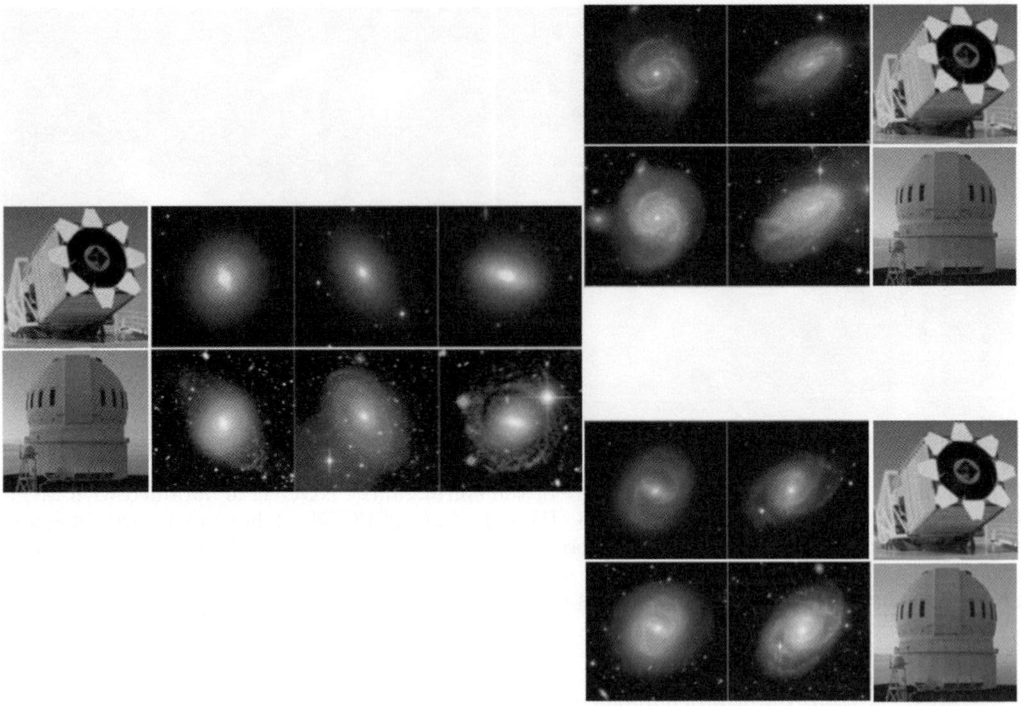

Figure 3. The Hubble diagram illustrated with images of 7 early and late type galaxies obtained with the Sloan (SDSS, top panels) and MegaCam at the CFHT. For the specific galaxies shown here, *the red and dead* ETGs turn out to exhibit a blue, star–forming, component on the the CFHT images, while the blue, star-forming, spirals appear to be embedded in a red stellar halo.

3.3. *Are fine structures present around all types of galaxies?*

Besides its depth, the originality of the MegaCam ATLAS^{3D} survey lies in the high number of observed galaxies: a complete sample of 260 nearby ETGs (plus about 100 spirals located in their field) when it is completed. This enables to get reliable statistical properties, to be compared with predictions from cosmological models of galaxy formation and associated numerical simulations. The latter predict that most, if not all, massive galaxies should be surrounded by stellar streams, i.e. the remnants of the multiple (mostly minor) mergers that presumably build them. The sensitivity limit of currently available surveys, including those presented here, and the fading of tidal debris with time allow us to detect the brightest streams, that trace relatively recent collisions (that occurred at redshift below 0.5–1).

Based on the catalog of 92 galaxies presented in Duc *et al.* (2014), we computed the fraction of galaxies showing signs of tidal perturbations. Galaxies were classified as on-going interacting systems, major mergers, minor mergers, and fully relaxed systems depending on the shape and number of fine structures around them. Preliminary results for two mass ranges are shown on Fig. 4. The galaxies in the high mass bin appear to have experienced more recent minor and major mergers. The fraction of fully relaxed systems drops by a factor of two from the low to the high mass bin. Similar trends are found going from fast to slow rotators, gas–poor to gas–rich objects, though selection criteria might still bias these results. Getting reliable statistics requires the completion of the survey.

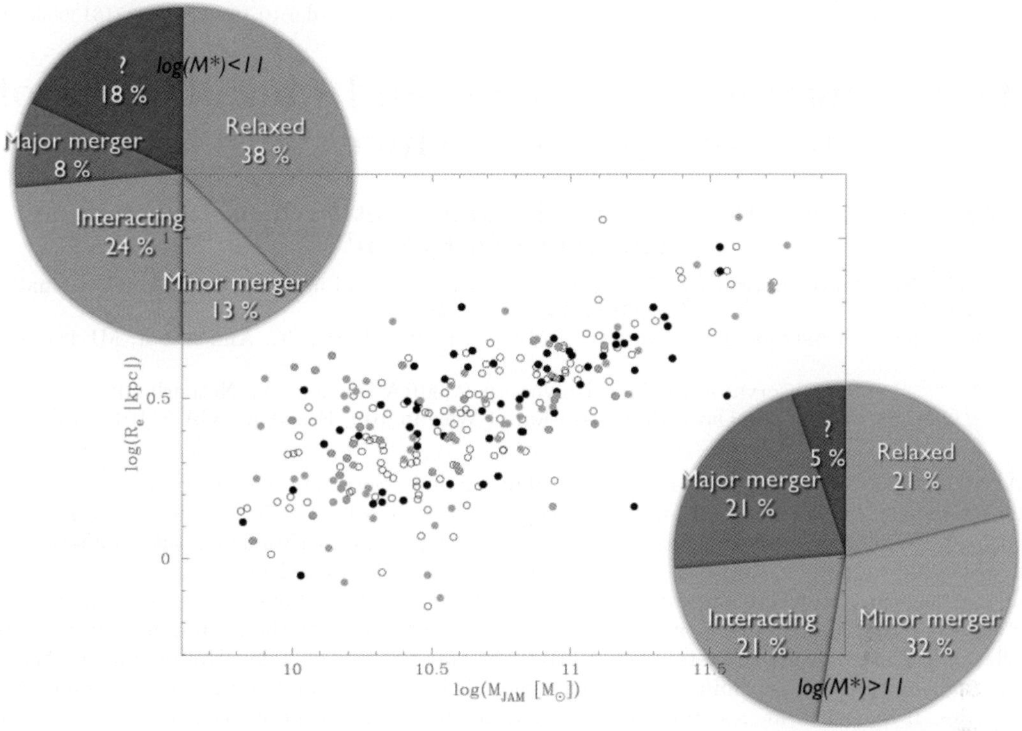

Figure 4. Dynamical mass, M_{JAM}, versus the effective radius, R_e (see Cappellari *et al.* 2013), of nearby early-type galaxies. Open circles correspond to the full ATLAS^{3D} sample, and filled discs are galaxies with available deep imaging. Relaxed galaxies showing no sign of tidal perturbations even with the deep imaging are shown in green. The cake diagrams on each side of the plot display the statistical results of the galaxy classification made with the deep images.

Acknowledgements

I am grateful to all my collaborators on this program made within the framework of the ATLAS^{3D}, NGVS and MATLAS projects. Many thanks in particular to Jean-Charles Cuillandre and Emin Karabal.

References

Cappellari, M., Scott, N., Alatalo, K., Blitz, L., Bois, M., *et al.* 2013, *MNRAS*, 432, 1709

Duc, P., Cuillandre, J., Karabal, E., *et al.* 2014, *MNRAS*, in press

Ferrarese, L., Côté, P., Cuillandre, J.-C., Gwyn, S. D. J., Peng, E. W., *et al.* 2012, *ApJS*, 200, 4

Kim, D. & Im, M., 2013, *ApJ*, 766, 109

Martínez-Delgado, D., Gabany, R. J., Crawford, K., Zibetti, S., & Majewski, S. R., 2010, *AJ*, 140, 962

Mihos, J. C., Harding, P., Feldmeier, J., & Morrison, H., 2005, *ApJ*, 631, L41

van Dokkum, P. G., Abraham, R., & Merritt, A., 2014, *ApJ*, 782, L24

York, D. G., Adelman, J., Anderson, Jr. J. E., Anderson, S. F., Annis, J., *et al.* 2000, *AJ*, 120, 1579

Galaxy Masses as Constraints of Formation Models
Proceedings IAU Symposium No. 311, 2014
M. Cappellari & S. Courteau, eds.

doi:10.1017/S174392131500352X

The evolution of the ages and metallicities of massive galaxies since $z = 0.7$

Anna Gallazzi[1], Eric F. Bell[2], Stefano Zibetti[1], Jarle Brinchmann[3] and Daniel D. Kelson[4]

[1]INAF-Osservatorio Astrofisico di Arcetri, Largo Enrico Fermi 5, 50125 Firenze, Italy. email: gallazzi@arcetri.astro.it
[2]Department of Astronomy, University of Michigan, 500 Church St., Ann Arbor, MI 48109, USA
[3]Leiden Observatory, Leiden University, 2300RA, Leiden, the Netherlands
[4]Observatories of the Carnegie Institution of Washington, Pasadena, CA 91101, USA

Abstract. We present results on the stellar population properties of massive galaxies at $z = 0.7$ based on deep, medium-resolution IMACS spectra for a sample of ~ 70 galaxies in the ECDFS with $M_* > 10^{10} M_\odot$. The age–mass and stellar metallicity–mass relations for the population as a whole have a similar shape as the local relations over the probed mass range, but offset to ages younger by ~ 4 Gyr and metallicities lower by $\sim$0.13 dex. Quiescent galaxies alone have stellar ages and metallicities consistent with passive evolution onto the local quiescent galaxies relations. The evolution in metallicity is driven by star-forming galaxies. However a significant fraction of massive star-forming galaxies have metallicities comparable to those of local quiescent galaxies. If quenched at $z < 0.7$ they can provide the necessary population to reproduce the scatter in age and metallicity of local quiescent galaxies.

1. Introduction

The population of massive galaxies has undergone a significant evolution since $z < 1$ as witnessed by the dramatic drop in the cosmic star formation rate (SFR) density (e.g. Cucciati *et al.* 2012) alongside a growth in the number density of quiescent galaxies (e.g. Moustakas *et al.* 2013; Muzzin *et al.* 2013). Tracing the ages and chemical abundances of the stellar populations in galaxies at different redshifts in relation to galaxy mass and star formation activity puts additional constraints on the mechanisms leading to the global suppression of star formation and the build-up of the quiescent galaxy population.

Studies of galaxy stellar populations at intermediate z are still limited as they require deep spectroscopy in the rest-frame optical where absorption features with distinct sensitivity to age and metallicity are located. Most of these works target red-sequence galaxies in clusters (e.g. Sánchez-Blázquez *et al.* 2009; Jørgensen & Chiboucas 2013) and find their properties to be consistent with passive evolution at high masses. So far only few works have presented the ages and metal abundances of massive quiescent galaxies in the field at $z \lesssim 0.7$ (Schiavon *et al.* 2006; Gallazzi *et al.* 2014; Choi *et al.* 2014).

Here we present results obtained on a sample of ~ 70 galaxies selected from the COMBO-17 catalog of the E-CDFS in the redshift range $0.65 < z < 0.75$ and with stellar mass $> 10^{10} M_\odot$. For this sample we obtained spectra with IMACS on the 6.5 m Magellan telescope at the Las Campanas Observatory covering the rest-frame range $3700 - 5500$Å with a spectral resolution of ~ 3.7Å rest-frame. As opposed to previous works we analyse both quiescent early-type galaxies and star-forming late-type galaxies. We quantify the relations between age, stellar metallicity and stellar mass for the population as a whole and for quiescent galaxies only and compare them with their analogs at

$z = 0$. By analysing individual galaxy spectra, we wish to understand whether individual galaxies that are already on the red-sequence at $z \sim 0.7$ undergo some evolution to $z = 0$ or whether the ensamble population evolves through addition of recently quenched galaxies, as suggested by number density evolution studies. In parallel, we analyse whether $z = 0.7$ star-forming galaxies have ages and stellar metallicities compatible with the population required for the observed red-sequence evolution.

2. Results

We estimate luminosity-weighted mean age, stellar metallicity and stellar mass following the bayesian approach developed in Gallazzi *et al.* (2005) to analyse SDSS galaxy spectra. We compare the strengths of the absorption features $D4000_n$, $H\beta$, $H\delta_A$+$H\gamma_A$, [Mg_2Fe], [MgFe]$'$ to a library of synthetic spectra based on the Bruzual & Charlot (2003) models convolved with a Monte Carlo library of star formation histories and metallicities.

In Fig. 1a we plot the median and the 16th and 84th percentiles of the age distribution in bins of stellar mass for our sample at $z = 0.7$ (filled stars with error bars). The mean stellar age increases from ~ 2.3 Gyr at $3 \times 10^{10} M_\odot$ to ~ 4.5 Gyr over an order of magnitude in mass. This can be directly compared with the analog age–mass distribution for SDSS galaxies (shaded area; Gallazzi *et al.* 2005): the shape of the relation at $z = 0.7$ is consistent within the uncertainties with that at $z = 0$, while the characteristic age at $10^{11.5} M_\odot$ is ~ 4 Gyr younger than the local value. The observed relation at $z = 0.7$ is significantly shallower and has a zero-point at an older age than what expected by passively evolving back in time the observed $z = 0$ relation (dotted curve). Similarly, in Fig. 1b the stellar metallicity distribution in bins of stellar mass for $z = 0.7$ galaxies is compared to the relation for SDSS galaxies. The shape of the relation is similar at both redshifts, while the characteristic stellar metallicity at $z = 0.7$ is ~ 0.13 dex lower than at $z = 0$. Both observations indicate that simple passive evolution of the entire massive galaxy population at $z = 0.7$ would not correctly predict the local age–mass and metallicity–mass relations.

We further separate galaxies into quiescent and star-forming based on a cut in specific SFR. The age and stellar metallicity distributions as a function of stellar mass of the $z = 0.7$ quiescent and star-forming galaxies are compared in Fig. 2 to the corresponding ones at $z = 0.1$ based on SDSS data. The ages of $z = 0.7$ quiescent galaxies display a similar increase with mass as local galaxies, but offset by ~ 3 Gyr. The stellar metallicities of $z = 0.7$ quiescent galaxies have a similar distribution as their $z = 0$ analogs over the same mass range, suggesting no evolution in stellar metallicity for this class of galaxies. The lack of evolution in the stellar metallicity distribution suggests that passive evolution applies to those massive galaxies that were already quiescent at intermediate redshift. Indeed their passively evolved ages are consistent with the local quiescent population (empty circles in Fig. 2a). However, the predicted scatter at $z = 0$ is lower than observed. The similar stellar metallicity as local galaxies and the low fraction of post-starburst galaxies (as we show in Gallazzi *et al.* 2014) allow for only a small amount of 'frosting' to slow down the aging of quiescent galaxies.

Star-forming galaxies have a steeper age–mass relation than quiescent galaxies, offset by ~ 5 Gyr from its local analog (Fig. 2c). Their stellar metallicities are on average ~ 0.12 dex lower than their local analogs. The observed offset between $z = 0.7$ and $z = 0$ is similar to the average evolution in gas-phase oxygen abundance of star-forming galaxies (e.g. Moustakas *et al.* 2011). It is also similar, although slightly larger, than the evolution predicted by Peeples & Somerville (2013) (0.07 dex) over the same mass and redshift range based on estimates of the average galaxy star formation histories and the

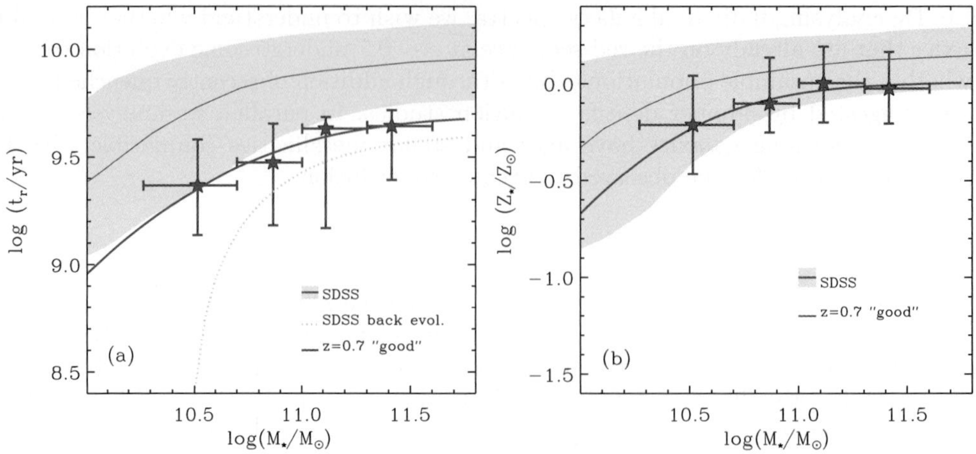

Figure 1. Luminosity-weighted mean age (panel *a*) and stellar metallicity (panel *b*) as a function of stellar mass for galaxies at $z = 0.7$. Filled stars show the median trends in bins of stellar mass, with error bars corresponding to the 16th and 84th percentiles. The solid curve shows the fitted relation. These relations are compared to the corresponding distributions for $z = 0.1$ SDSS galaxies. The dotted curve in panel *a* indicates the age–mass relation at $z = 0.7$ expected by passive backward evolution of the SDSS relation. Adapted from Fig. 7 of Gallazzi *et al.* (2014).

observed mass-gas metallicity-SFR relation. The observed evolution in the average stellar metallicity indicates chemical enrichment in at least some of the massive star-forming galaxies. We note though that a significant fraction of $z = 0.7$ star-forming galaxies have metallicities already comparable to those of local quiescent galaxies. These galaxies have stellar population properties compatible with those required to populate the younger part of the local quiescent galaxy population, increasing the scatter in age without altering the metallicity distribution.

3. Conclusions

Based on deep, medium-resolution IMACS spectroscopy of a sample of ~ 70 massive galaxies at $z = 0.7$, we characterize for the first time at these redshifts the stellar metallicity and age scaling relations for the population as a whole and for quiescent and star-forming galaxies separately We find that $z = 0.7$ quiescent galaxies have stellar metallicities and stellar ages consistent with local quiescent galaxies under passive evolution hypothesis, in qualitative agreement with results on cluster quiescent galaxies by Sánchez-Blázquez *et al.* (2009) and Jørgensen & Chiboucas (2013). However, we also show that the scatter in age of $z = 0.7$ quiescent galaxies is too low to reproduce the scatter of the local relation. Addition of recently-quenched star-forming galaxies could contribute to the build-up of the younger portion of the local red-sequence. This possibility was also suggested for field galaxies by Schiavon *et al.* (2006) and is not ruled out by Choi *et al.* (2014). About 40% of massive star-forming galaxies in our sample have indeed stellar ages and metallicities compatible with those required to increase the scatter in age without affecting the metallicity distribution to reproduce the local quiescent population. Upcoming large deep spectroscopic programs will allow to extend these studies to lower stellar masses and to better sample the star-forming galaxy population. Comparison of these studies with those based on cluster galaxies can shed light on the role of environment in setting the mass scale where galaxies evolve passively.

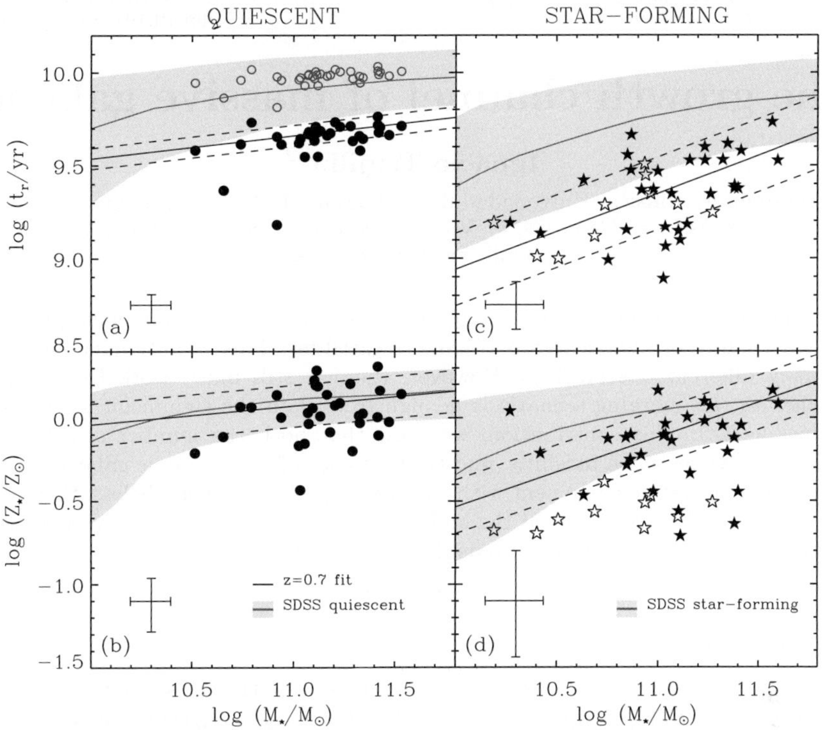

Figure 2. Luminosity-weighted mean age (panels *a* and *b*) and stellar metallicity (panels *c* and *d*) versus stellar mass for quiescent (*left*) and star-forming galaxies (*right*; empty stars indicate $z = 0.7$ galaxies with poor metallicity constraints). The shaded regions and solid curves show the relations for SDSS galaxies, while the solid and dashed lines show the linear relation and scatter fitted to $z = 0.7$ galaxies. The empty circles in panel *a* indicate the expected location on the local relation of $z = 0.7$ quiescent galaxies if they evolved passively until $z = 0.1$. The error bars in each panel represent the mean error on the parameters for each subsample of $z = 0.7$ galaxies. Adapted from Fig. 12 of Gallazzi *et al.* (2014).

Acknowledgements

AG acknowledges funding from the European Union FP7/2007-2013 under grant agreement n. 267251 (AstroFIt). This work is based on data gathered with the 6.5 meter Magellan Telescopes located at Las Campanas Observatory, Chile.

References

Bruzual, G. & Charlot, S., 2003, *MNRAS*, 344, 1000
Choi, J., Conroy, C., Moustakas, J., *et al.* 2014, *ApJ*, 792, 95
Cucciati, O., Tresse, L., Ilbert, O., *et al.* 2012, *A&A*, 539, A31
Gallazzi, A., Charlot, S., Brinchmann, J., White, S. D. M., & Tremonti, C. A. 2005, *MNRAS*, 362, 41
Gallazzi, A., Bell, E. F., Zibetti, S., Brinchmann, J., & Kelson, D. D. 2014, *ApJ*, 788, 72
Jørgensen, I. & Chiboucas, K. 2013, *AJ*, 145, 77
Moustakas, J., Zaritsky, D., Brown, M., *et al.* 2011, arXiv:1112.3300
Moustakas, J., Coil, A. L., Aird, J., *et al.* 2013, *ApJ*, 767, 50
Muzzin, A., Marchesini, D., Stefanon, M., *et al.* 2013, *ApJ*, 777, 18
Peeples, M. S. & Somerville, R. S. 2013, *MNRAS*, 428, 1766
Sánchez-Blázquez, P., Jablonka, P., Noll, S., *et al.* 2009, *A&A*, 499, 47
Schiavon, R. P., Faber, S. M., Konidaris, N., *et al.* 2006, *ApJ*, 651, L93

Galaxy Masses as Constraints of Formation Models
Proceedings IAU Symposium No. 311, 2014
M. Cappellari & S. Courteau, eds.

doi:10.1017/S1743921315003531

The growth channel of massive galaxies

Ignacio Trujillo[1,2]

[1]Departamento de Astrofísica, Universidad de La Laguna, E-38205, La Laguna, Tenerife, Spain
[2]Instituto de Astrofísica de Canarias, c/ Vía Láctea s/n, E-38205, La Laguna, Tenerife, Spain
email: `trujillo@iac.es`

Abstract. There is growing evidence suggesting that massive galaxies have growth both in mass and size mainly by the accretion of smaller satellites. This evolutionary path qualitative explains many different observations. However, there is still much work to do on trying to quantify whether this merging scenario is enough to explain the significant growth of massive galaxies since z$\sim$2. In this contribution, we show that both the number of satellite galaxies around massive galaxies at all redshifts, and the existence of massive relic galaxies in the nearby Universe are in quantitative agreement with this merging picture. Nonetheless, there is still open questions related to the properties of the stellar halos of present-day massive galaxies and the number density of massive relics at all redshifts that remain open.

1. Introduction

How did the enormous cosmic structural evolution of the most massive galaxies take place (Daddi *et al.* 2005; Trujillo *et al.* 2006)? There is a currently growing consensus that the size evolution (around a factor of 4 since z$\sim$2; Trujillo *et al.* 2007; Buitrago *et al.* 2008) was driven by a continuous accretion of satellites towards the main objects (e.g. Bezanson *et al.* 2009; Hopkins *et al.* 2009; Oser *et al.* 2010). However, although *in-situ* growth scenarios (Fan *et al.* 2008; Ragone-Figueroa & Granato 2011) have been disfavored observationally (see a full discussion in e.g. Trujillo, Ferreras, & de La Rosa 2011), *ex-situ* models (i.e. merging scenarios Ciotti & van Albada 2001; Khochfar & Burkert 2006; Naab, Johansson, & Ostriker 2009) lack a detailed *quantitative* confrontation with the observational data. What follows are two observational tests that merging scenarios have to observationally pass to be considered as a valid growth channel:

- Have we got enough number of mergers to produce the size increase?
- Where are the untouched massive *relic* galaxies in the nearby Universe?

We will expand on these questions in the following sections.

2. Have we got enough number of mergers to produce the size increase?

If the merging channel is the predominant path that massive galaxies have followed to reach their current sizes, at all redshifts there should be enough satellites surrounding the massive galaxies to produce, by accretion, the observed size evolution. There is an increasing number of papers exploring the frequency of satellites around the most massive galaxies at different cosmic epochs (e.g. Kaviraj *et al.* 2009; Bundy *et al.* 2009; Mármol-Queraltó *et al.* 2012; López-Sanjuan *et al.* 2012; Ruiz, Trujillo, & Mármol-Queraltó 2014). The strategy followed in those works consists on counting the number of galaxies within a given radial distance and redshift interval to the targeted massive galaxies. This is illustrated in Fig. 1. The amount of satellites around massive galaxies have remained pretty much constant with cosmic time at least since z$\sim$1.5. This is quite independent of the mass ratio between the host and the satellite galaxy. This is shown in Fig. 2.

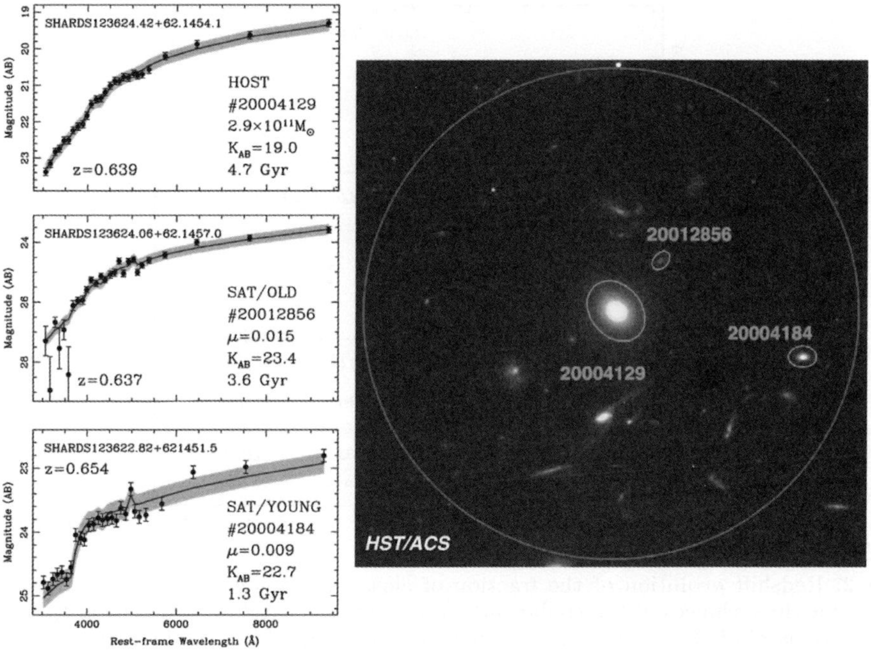

Figure 1. Illustration of how satellites are identified around massive galaxies (figure taken from Ferreras *et al.* 2014). The host is the massive galaxy SHARDS123624.42+621454 (z=0.639, $M_\star=2.9\times10^{11}$ $M_\odot$). The 30×30 arcsec^2 image (right) is an RGB colour composite from the GOODS-N HST/ACS archival data through the F606W, F775W, and F850LP passbands. The red circumference extends over the $100h_{70}^{-1}$ kpc search radius. On the left, the observed data points (black circles) are compared with the best fitting models (red lines). Each galaxy is labeled by its SHARDS ID, K_{AB} apparent magnitude, best-fitting stellar age, and either stellar mass (host) or mass ratio (companion).

A detailed analysis of the mass contribution of each satellite to the whole system host-satellites shows that, in case of infall of the satellites to the galaxy host, the largest contributor to the mass growth of the host would be those satellites with the largest masses (Ruiz, Trujillo, & Mármol-Queraltó 2014; Ferreras *et al.* 2014). This is because there are not enough number of small satellites to balance the contribution of the most massive satellites.

In Fig. 3 we show what would be the average mass growth of the massive galaxies assuming two different infall timescale of the satellites to the host galaxies. The first one assumes that the infall time scale is independent of the mass of the satellite. The other one takes into account the dynamical friction (Jiang, Jing, & Han 2014), and consequently, considers more likely that the most massive satellites have shorter infall time scales. Independently, of the exact details of the infall timescale, the average mass growth of the massive galaxies due to the infall of satellites is ∼8% per Gyr. This translates into a global mass growth since z∼1 of approximately a factor of 2.

According to the literature, the growth in size is connected to the growth in mass by simple scaling relationships: $\Delta R \propto \Delta M^{\alpha}$ with $1 \lesssim \alpha \lesssim 2$. Consequently, an increase in mass of ∼2 will correspond to an increase in size of ∼2-4. This is more than enough to explain the observed size evolution of the massive galaxies since z∼1, which is around a factor of 2 (e.g. Trujillo *et al.* 2007). In this sense, the merger channel seems to have enough number of satellites around the massive galaxies to be able to explain the observed size growth without significant problems.

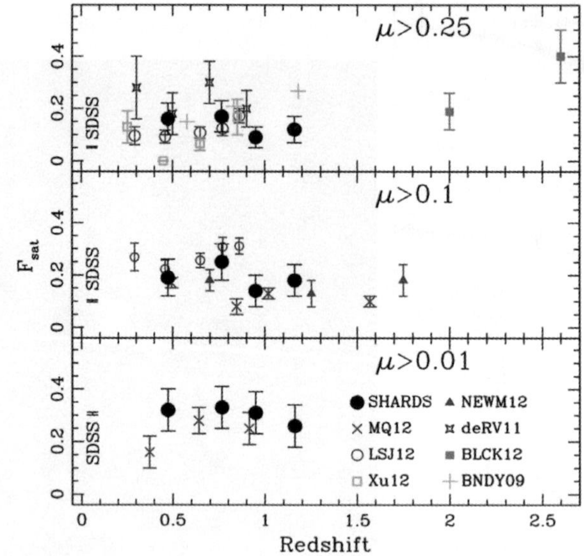

Figure 2. Redshift evolution of the fraction of close companions of massive galaxies ($M_\star > 10^{11} M_\odot$), for three choices of the stellar mass ratio between satellite and central (μ). The solid circles (Ferreras *et al.* 2014) are compared with recent results from the literature.

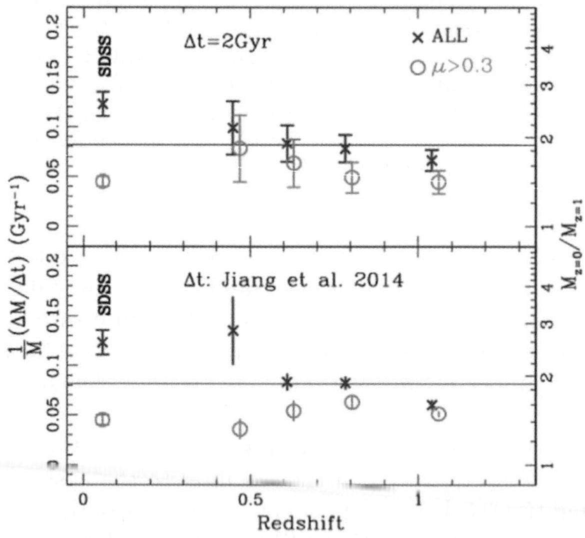

Figure 3. Top: growth rate of massive galaxies with respect to redshift for the simple case of a fixed (Δt=2 Gyr) merging time-scale. The red crosses (grey circles) correspond to the contribution from the whole sample (only from major mergers, μ >0.3). The axis on the right gives the expected stellar mass growth between z=1 and 0 assuming a constant rate. For reference, the points on the left, labelled SDSS, correspond to a nearby sample. Bottom: same as above, although taking into account the dependence of the merging time-scale on projected separation and mass ratio, according to the prescription of (Jiang, Jing, & Han 2014).

3. Where are the untouched massive *relic* galaxies in the nearby Universe?

The following question we want to address is whether there is any single massive nearby galaxy in our vicinity such it has remained unevolved since its early formation ($z_f \gtrsim 2$).

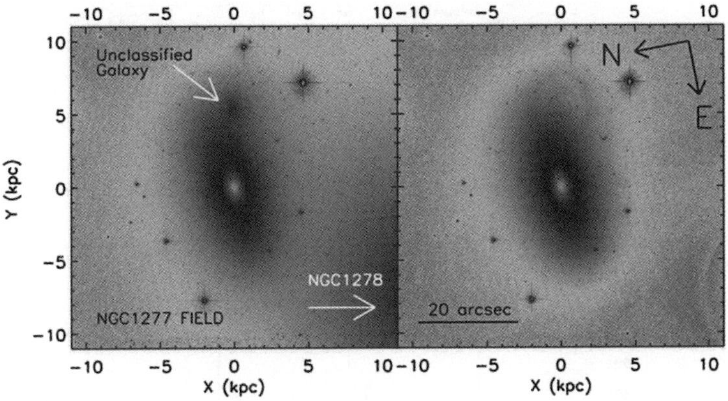

Figure 4. Neighborhood of the massive relic galaxy NGC 1277 as seen by the HST F625W filter. The left panel shows the two closest galaxies whose light contaminate NGC 1277. The right panel shows NGC 1277 after the subtraction of the contaminant light. The results indicates that NGC 1277 is rather symmetric with no distortions or bright tidal streams surrounding it.

A recent work (Quilis & Trujillo 2013), using the Millenium simulation, suggests that 1 every 1000 massive ($M_\star > 10^{11} M_\odot$) galaxies in the present-day universe has accreted less than 10% of its total stellar mass since z=2. Taking into account the number density of massive galaxies today, this is equivalent to find around one massive relic galaxy every 10^6 Mpc^3 (i.e. one in every cosmic sphere of radius ~63 Mpc).

Detailed analysis of our vicinity (Trujillo *et al.* 2009; Taylor *et al.* 2010; Poggianti *et al.* 2013) have failed to find any single massive galaxy with characteristics of a relic galaxy (i.e. $M_\star > 10^{11} M_\odot$, $R_e < 1.5$ kpc and with stellar populations homogeneously old, age > 10 Gyr, along its whole structure). However, a very nearby object, NGC1277, not included in the catalogues previously used, presents structural features that makes it very attractive to be considered as a potential massive relic galaxy. In fact, its global properties ($M_\star = 1.2\times10^{11} M_\odot$, $R_e = 1.2$ kpc, $\sigma > 330$ km/s; van den Bosch *et al.* 2012) are within the expected properties of a galaxy with relic characteristics. We show in Fig. 4 the morphology of this galaxy located in the rich Perseus Cluster of galaxies at 73 Mpc. According to its regular morphology, NGC1277 does not present signs of currently (or in its immediate past) accreting external stars.

Not only the structural properties of NGC1277 are consistent with the structural properties of massive galaxies at z~2 but also its stellar populations are old along its radial structure (Trujillo *et al.* 2014) as expected if this object were a truly relic. In Fig. 5 we show the spectral energy distribution of the galaxy at different radial distances based on very deep spectra taken with the WHT telescope. At all radii, the Star Formation Histories (SFHs) of this galaxy is compatible with no new star formation in the last 10 Gyr.

Having identified a good candidate of a relic massive galaxy in the nearby Universe, this opens the possibility of exploring with exquisite detail the properties of the early steps of massive galaxy formation. In Fig. 6 we show the age, metallicity and α/Fe radial profiles of NGC1277. The large α/Fe value, α/Fe>0.3, implies an extremely short formation time-scale: ~100 Myr. This short time interval translates into a huge star formation rate to form the bulk of NGC1277: ~1000 $M_\odot$/yr.

To conclude this section, the existence of NGC1277 strongly suggests that some massive galaxies have remained unaltered since z~2. This is in qualitative agreement with merging scenarios where some of these objects are expected to survive. These relic galaxies, on

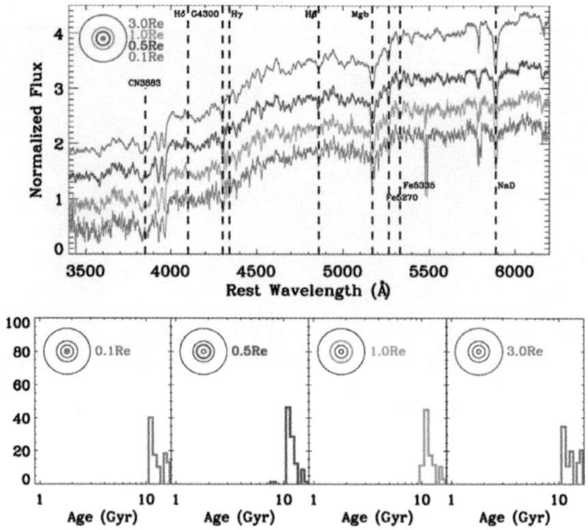

Figure 5. Spectral energy distribution of NGC1277 at different radial distances (Trujillo *et al.* 2014). The spectra are shown in arbitrary units, with the flux normalized per unit wavelength and shifted for clarity. The position of several relevant absorption lines are indicated with vertical lines. The bottom panels show the SFHs derived with STARLIGHT at different radial distances. This represents the fraction of mass created at each epoch.

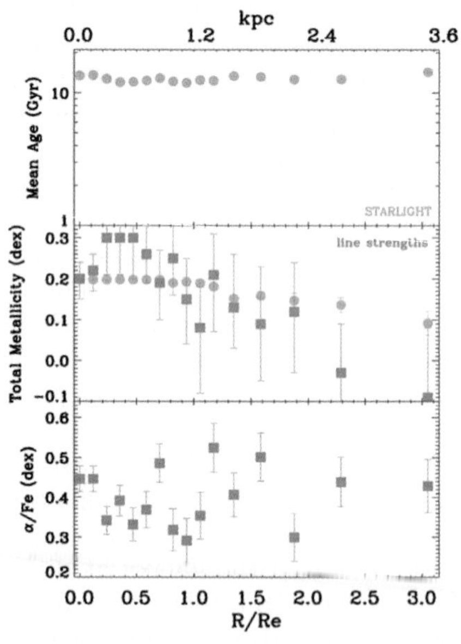

Figure 6. Age, metallicity, and α/Fe profiles of NGC 1277 (Trujillo *et al.* 2014). The figure shows the mean mass-weighted age and total metallicity derived using STARLIGHT. An independent measure of the metallicity and α/Fe was obtained from an indices analysis. The profiles show a small change of the stellar population properties of this relic galaxy across its structure out to $3R_e$. It seems that the entire galaxy was formed in a unique, very fast event, which produced the high α/Fe abundances.

the other hand, are difficult to understand in growing scenarios driven by AGN activity, as NGC1277 already hosts a very large supermassive blackhole (van den Bosch *et al.* 2012) and still remains compact.

4. Conclusions

The merging channel mode of galaxy growth successfully passes the tests: a) of having enough number of satellites around the hosts to explain the size and mass growth, and b) it also explains the existence of unevolved massive galaxies in the nearby Universe. Future works would need to explore whether the number density of unevolved massive galaxies at all redshifts is in agreement with cosmological predictions (see e.g. Damjanov *et al.* 2014). In addition, it will be necessary to explore whether the outskirts of nearby massive galaxies have the same properties (age and metallicity) than the satellites found at high-z (argueably the most likely progenitors of these envelopes).

Acknowledgements

This work has been supported by the "Programa Nacional de Astronomía y Astrofísica" of the Spanish Ministry of Science and Innovation under grant AYA2010-21322-C03-02.

References

Bezanson, R., van Dokkum, P. G., Tal, T., Marchesini, D., Kriek, M., Franx, M., & Coppi, P., 2009, *ApJ*, 697, 1290

Buitrago, F., Trujillo, I., Conselice, C. J., Bouwens, R. J., Dickinson, M., & Yan, H., 2008, *ApJ*, 687, L61

Bundy, K., Fukugita, M., Ellis, R. S., Targett, T. A., Belli, S., & Kodama, T., 2009, *ApJ*, 697, 1369

Ciotti, L. & van Albada, T. S., 2001, *ApJ*, 552, L13

Daddi, E., *et al.*, 2005, *ApJ*, 626, 680

Damjanov, I., Hwang, H. S., Geller, M. J., & Chilingarian, I., 2014, *ApJ*, 793, 39

Fan, L., Lapi, A., De Zotti, G., & Danese, L., 2008, *ApJ*, 689, L101

Ferreras, I., *et al.*, 2014, *MNRAS*, 444, 906

Hopkins, P. F., Bundy, K., Murray, N., Quataert, E., Lauer, T. R., & Ma, C.-P., 2009, *MNRAS*, 398, 898

Jiang, C. Y., Jing, Y. P., & Han, J., 2014, *ApJ*, 790, 7

Kaviraj, S., Peirani, S., Khochfar, S., Silk, J., & Kay, S., 2009, *MNRAS*, 394, 1713

Khochfar, S. & Burkert, A., 2006, *A&A*, 445, 403

López-Sanjuan, C., *et al.*, 2012, *A&A*, 548, A7

Mármol-Queraltó, E., Trujillo, I., Pérez-González, P. G., Varela, J., & Barro, G., 2012, *MNRAS*, 422, 2187

Naab, T., Johansson, P. H., & Ostriker, J. P., 2009, *ApJ*, 699, L178

Oser, L., Ostriker, J. P., Naab, T., Johansson, P. H., & Burkert, A., 2010, *ApJ*, 725, 2312

Poggianti, B. M., *et al.*, 2013, *ApJ*, 762, 77

Quilis, V. & Trujillo, I., 2013, *ApJ*, 773, L8

Ragone-Figueroa, C. & Granato, G. L., 2011, *MNRAS*, 414, 3690

Ruiz, P., Trujillo, I., & Mármol-Queraltó, E., 2014, *MNRAS*, 442, 347

Taylor, E. N., Franx, M., Glazebrook, K., Brinchmann, J., van der Wel, A., & van Dokkum, P. G., 2010, *ApJ*, 720, 723

Trujillo, I., *et al.*, 2006, *MNRAS*, 373, L36

Trujillo, I., Conselice, C. J., Bundy, K., Cooper, M. C., Eisenhardt, P., & Ellis, R. S., 2007, *MNRAS*, 382, 109

Trujillo, I., Cenarro, A. J., de Lorenzo-Cáceres, A., Vazdekis, A., de la Rosa, I. G., & Cava, A., 2009, *ApJ*, 692, L118

Trujillo, I., Ferreras, I., & de La Rosa, I. G., 2011, *MNRAS*, 415, 3903

Trujillo, I., Ferré-Mateu, A., Balcells, M., Vazdekis, A., & Sánchez-Blázquez, P., 2014, *ApJ*, 780, L20

van den Bosch, R. C. E., Gebhardt, K., Gültekin, K., van de Ven, G., van der Wel, A., & Walsh, J. L., 2012, Natur, 491, 729

van Dokkum, P. G., *et al.*, 2010, *ApJ*, 709, 1018

Galaxy Masses as Constraints of Formation Models
Proceedings IAU Symposium No. 311, 2014
M. Cappellari & S. Courteau, eds.

doi:10.1017/S1743921315003543

Initial Mass Function for Massive Galaxies at $z \sim 1$

Shravan Shetty[1] **and Michele Cappellari**

Sub-department of Astrophysics, Department of Physics, University of Oxford, Denys Wilkinson Building, Keble Road, Oxford OX1 3RH.
[1]Email: shravan.shetty@astro.ox.ac.uk

Abstract. We present the results on the stellar Initial Mass Function (IMF) normalisation of 68 massive ($M_* = 10^{11} - 10^{12} M_\odot$) Early-Type Galaxies (ETGs) at redshift of ∼1. This was achieved by deriving the stellar Mass-to-Light ratio (M/L) of the galaxies through axis-symmetric dynamical modelling and comparing it to the same derived via stellar population modelling through full spectrum fitting. The study also employs an Abundance Matching technique to account for the dark matter within the galaxies. The results demonstrate that massive ETGs at high redshifts on average have a Salpeter-like IMF normalisation, while providing observational evidence supporting previous predictions of low dark matter fraction in the inner regions ($<1R_e$) of galaxies at higher redshift.

Keywords. galaxies: evolution, galaxies: high-redshift, galaxies: star formation, galaxies: stellar content

1. Introduction

In recent years, a significant paradigm shift has taken place in our understanding of the Initial Mass Function (IMF) of galaxies. In Cappellari *et al.* (2013a, Hereafter ATLAS^{3D}XV), the authors created detailed axis-symmetric dynamical models to robustly derive the stellar Mass-to-Light ratio (M/L) for 260 ETGs in the local universe. By comparing these M/Ls to those derived through stellar population modelling, the authors demonstrated that the IMF of galaxies wasn't universal, but in fact varied systematically with the central velocity dispersion of the galaxies (Cappellari *et al.* 2012, 2013b, Hereafter ATLAS^{3D}XX). This systematic variation has been confirmed via stellar population models (Spiniello *et al.* 2012; Conroy & van Dokkum 2012; Smith *et al.* 2012; Ferré-Mateu *et al.* 2013; La Barbera *et al.* 2013) and dynamical approaches (Dutton *et al.* 2013; Tortora *et al.* 2013). In this work, we have attempted to study the IMF normalisation of high redshift galaxies.

2. Data

We use 1D galaxy spectra taken from the DEEP2 Galaxy Redshift Survey (Newman *et al.* 2013). The survey has observed ∼49,000 galaxies in an observed-frame wavelength range of 6,500-9,100Å at a resolution of R∼6,000 at 7,800Å. We also use high-resolution (0.03"/pixel) F814W filter HST/ACS images of our galaxies, taken from the AEGIS data product (Davis *et al.* 2007), which translates to the rest-frame B-band for our galaxies.

3. Method

Before we start our analysis we apply our first set of selection criteria. We restrict our analysis to galaxies in the DEEP2 survey that have reliable redshift and lay in the

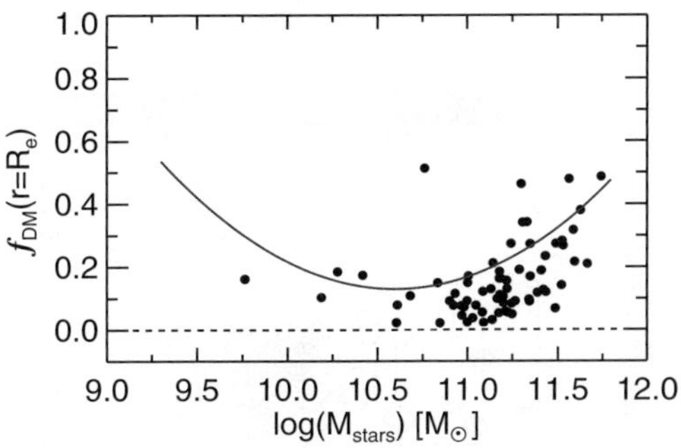

Figure 1. This plot depicts the dark matter fraction within $1R_e$ of our galaxies and it's systemmatic variation with galaxy stellar mass. The solid line represents the trend found by ATLAS^{3D}XV (Bottom panel of fig. 10). The plot above is directly comparable to that in ATLAS^{3D}XV since both were produced using a similar Abundance matching technique.

redshift range of 0.7-0.9. Along with this, we also visually inspect the galaxy spectra to verify the presence of significant absorption lines, and to mask telluric features and significant gas emission lines.

We derive the stellar kinematics of the galaxies by fitting the galaxy spectra to a subset of 53 empirical stellar spectra taken from the Indo-US Library of Coudé Feed Stellar Spectra Library (Valdes *et al.* 2004), using the pPXF code (Cappellari & Emsellem 2004). We have set a Signal-to-Noise (S/N) cutoff at 3 per 60 km/s spectral pixel and have visually inspected every spectrum to ensure that the best-fits do indeed fit real stellar features. We have derived realistic errors using a bootstrapping technique.

To create our dynamical models, we use the JAM dynamical modelling code which solves the anisotropic Jean's equations under the assumption of axis-symmetry (Cappellari 2008), while taking into account aperture and seeing effects. Using the light and mass profile for the galaxy, the code is able to predict the expected observed stellar kinematics for the galaxy profiles which can then be calibrated with the observed kinematics to derive the stellar M/L of the galaxies.

To parametrise the light and stellar mass profile of our galaxies, we use the MGE fitting code (Cappellari 2002) on our galaxy photometry. For our galaxies, it is essential that we accurately reproduce the light profile of the inner region of galaxies, where the stellar kinematics are derived, and hence we have visually verified each galaxy fit. In the process of visual verification, we have eliminated galaxies that have non-axis-symmetric features such as bars, dust lanes etc as these aren't well parametrised by the code.

To model the halo profile, we have used results from Abundance Matching. To do this, we associate every possible stellar mass, or stellar M/L, of the galaxy with an associated halo mass at the given redshift of the galaxy using the results of Moster *et al.* (2013). Assuming that the dark matter halo follows a NFW profile (Navarro *et al.* 1996), we use the results of Klypin *et al.* (2011) to derive the concentration of the profile, while accounting for the redshift of the galaxies, and thereafter other physical parameters of the halo. We can then derive the best fitting stellar mass, ie stellar M/L, which along with it's associated dark matter halo best reproduces the observed stellar kinematics of our galaxies through our dynamical modelling.

To use the JAM code we have to make certain assumptions on our galaxies, ie. we assume a constant velocity anisotropy ($\beta_z \equiv 1 - \sigma_z^2/\sigma_R^2$) of 0.2 for our galaxies, and assume that the inclination of our galaxies is 60° unless this is unrealistic given the galaxy photometry. These assumptions are supported by results of realistic simulations carried out by Lablanche *et al.* (2012) and previous works (Cappellari *et al.* 2006; Cappellari

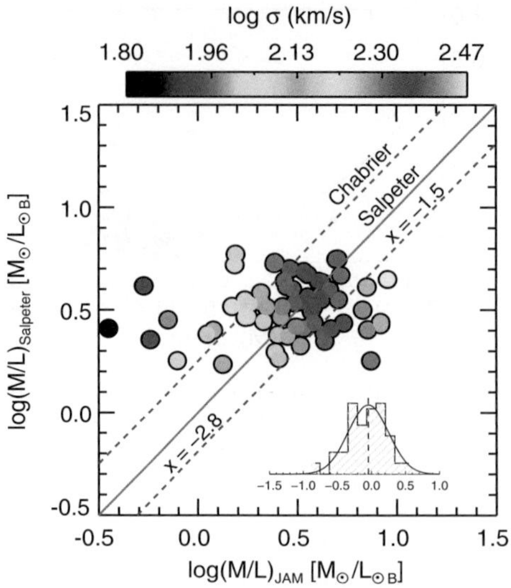

Figure 2. This plot illustrates the IMF normalisation of the galaxies. The x-axis represents the stellar M/L derived via our dynamical models, while the y-axis represents that derived through stellar population modelling. The plot is is directly comparable to fig.11 in ATLAS^{3D}XX.

2008) that have shown that the errors introduced by the uncertainty in these quantities are insignificant compared to the errors on the stellar kinematics of our galaxies.

Next, we derive the stellar M/Ls through full spectrum fitting of the galaxy spectra with a template model set using pPXF. We use the MILES stellar population models (Vazdekis *et al.* 2010), which are based high resolution empirical stellar spectra of the MILES stellar library (Sánchez-Blázquez *et al.* 2006), under the assumption of a Salpeter IMF. We also assume that the star formation history of the galaxies varies linearly in log-time space.

To implement our final selection criterion, we visually inspect the stellar populations of our galaxies and remove those that contain multiple major star formation events or significant fraction of young (<1.2Gyr) stellar population. This criterion is based on results of ATLAS^{3D}XV where the authors find that such young galaxies tend to have a radially varying stellar M/L, and hence break the implicit assumption of constant M/L throughout the galaxy.

For further information on the methodology used and examples plots of the various fits in the work, we refer the reader to Shetty & Cappellari (2014), where the same analysis is done with the exception of Abundance Matching.

4. Results and Conclusion

The work presented here extends to higher redshift earlier works of the IMF normalization of local galaxies. Through a rigorous selection criteria, we have carefully derived the stellar M/L for 68 galaxies, from an original sample of ∼49,000 galaxies, through axis-symmetric dynamical modelling and stellar population modelling. As a result of these selection criteria, our galaxy sample is biased towards massive ($M_* = 10^{11} - 10^{12}\ M_\odot$) ETGs.

The dynamical M/L derived in this work represents the stellar M/L of the galaxy as it takes into account the contribution of the dark matter halo through Abundance Matching. Based on the results of detailed dynamical modelling in ATLAS^{3D}XV and on merger computer simulations of Hilz *et al.* (2013), Shetty & Cappellari (2014) had assumed that the dark matter fraction within the inner $1R_e$ of these galaxies is <30%. The abundance matching technique used in this study supports this prediction. In Fig. 1, we depict the dark matter fraction of our galaxies. Here we notice a trend of the dark matter fraction with the stellar mass of the galaxies, which is similar to that observed by ATLAS^{3D}XV (Bottom panel of fig. 10), with a lower normalisation, consistent with the results of Hilz *et al.* (2013).

In Fig. 2, we present the IMF normalisation of our galaxies. The colour code in the plot represents the smoothed variation of the central velocity dispersion of the galaxies. The plot clearly illustrates that, on average, the IMF normalisation of the massive high redshift ETGs is Salpeter-like. This plot also qualitatively similar to fig. 11 of ATLAS^{3D}XX, ie both plot illustrate a lack of correlation between the two stellar M/Ls for the most massive galaxies. Assuming the passive evolution of the central regions of these galaxies, this result is consistent with results produced by ATLAS^{3D}, Auger *et al.* (2010, SLACS), Conroy (2013) and Spiniello *et al.* (2014).

References

Auger, M. W., Treu, T., Gavazzi, R., *et al.* 2010, *ApJ*, 721, L163
Cappellari, M. 2002, *MNRAS*, 333, 400
Cappellari, M. 2008, *MNRAS*, 390, 71
Cappellari, M. & Emsellem, E. 2004, *PASP*, 116, 138
Cappellari, M., Bacon, R., Bureau, M., *et al.* 2006, *MNRAS*, 366, 1126
Cappellari, M., McDermid, R. M., Alatalo, K., *et al.* 2012, *Nature*, 484, 485
Cappellari, M., Scott, N., Alatalo, K., *et al.* 2013a, *MNRAS*, 432, 1709 (ATLAS^{3D}XV)
Cappellari, M., McDermid, R. M., Alatalo, K., *et al.* 2013b, *MNRAS*, 432, 1862 (ATLAS^{3D}XX)
Conroy, C. 2013, *ARA&A*, 51, 393
Conroy, C. & van Dokkum, P. G. 2012, *ApJ*, 760, 71
Davis, M., Guhathakurta, P., Konidaris, N. P., *et al.* 2007, *ApJ*, 660, L1
Dutton, A. A., Macciò, A. V., Mendel, J. T., & Simard, L. 2013, *MNRAS*, 432, 2496
Ferré-Mateu, A., Vazdekis, A., & de la Rosa, I. G. 2013, *MNRAS*, 431, 440
Hilz, M., Naab, T., & Ostriker, J. P. 2013, *MNRAS*, 429, 2924
Klypin, A. A., Trujillo-Gomez, S., & Primack, J. 2011, *ApJ*, 740, 102
La Barbera, F., Ferreras, I., Vazdekis, A., *et al.* 2013, *MNRAS*, 433, 3017
Lablanche, P.-Y., Cappellari, M., Emsellem, E., *et al.* 2012, *MNRAS*, 424, 1495
Moster, B. P., Naab, T., & White, S. D. M. 2013, *MNRAS*, 428, 3121
Navarro, J. F., Frenk, C. S., & White, S. D. M. 1996, *ApJ*, 462, 563
Newman, J. A., Cooper, M. C., Davis, M., *et al.* 2013, *ApJS*, 208, 5
Sánchez-Blázquez, P., Peletier, R. F., Jiménez-Vicente, J., *et al.* 2006, *MNRAS*, 371, 703
Shetty, S. & Cappellari, M. 2014, *ApJ*, 786, L10
Smith, R. J., Lucey, J. R., & Carter, D. 2012, *MNRAS*, 426, 2994
Spiniello, C., Trager, S., Koopmans, L. V. E., & Conroy, C. 2014, *MNRAS*, 438, 1483
Spiniello, C., Trager, S. C., Koopmans, L. V. E., & Chen, Y. P. 2012, *ApJ*, 753, L32
Tortora, C., Romanowsky, A. J., & Napolitano, N. R. 2013, *ApJ*, 765, 8
Valdes, F., Gupta, R., Rose, J. A., Singh, H. P., & Bell, D. J. 2004, *ApJS*, 152, 251
Vazdekis, A., Sánchez-Blázquez, P., Falcón-Barroso, J., *et al.* 2010, *MNRAS*, 404, 1639

Galaxy Masses as Constraints of Formation Models
Proceedings IAU Symposium No. 311, 2014
M. Cappellari & S. Courteau, eds.

doi:10.1017/S1743921315003555

The Secret Life of Galaxies

Alan Dressler[1] and Louis Abramson[2]

[1]The Observatories of the Carnegie Institution of Science, 813 Santa Barbara Street, Pasadena, CA 91101-1232, USA. email: dressler@obs.carnegiescience.edu
[2]Department of Astronomy & Astrophysics and Kavli Institute for Cosmological Physics, The University of Chicago, 5640 South Ellis Avenue, Chicago, IL 60637, USA. email: labramson@uchicago.edu>

Abstract. We have learned much about galaxy evolution since z = 2, and something to even higher redshifts. How can it be that we know so little about the star formation histories (SFHs) of individual galaxies? Although great progress has been made accumulating huge samples with only rudimentary properties, progress in galaxy evolution means connecting what we've learned to detailed measurements of the life-histories of specific — not just representative — systems.

Keywords. galaxies: evolution - galaxies: formation

1. Introduction

After a decade of scatter diagrams showing a point for each of hundreds of thousands of galaxies, each measured in some simple way that belies the complexity of the complicated life-histories that each galaxy, there is renewed interest in connecting the detailed knowledge we have about the Milky Way and its neighbors with the billions of galaxies we see at past epochs, almost to their very beginnings. In this paper we present examples of this shift from the general to the particular that promise to move the field forward.

2. The Green Valley is Purple

The Sloan Digital Sky Survey (SDSS) provided, for the first time, photometry for hundreds of thousands of nearby galaxies. Among the first remarkable results in galaxy evolution to be addressed with this enormous data sample was the issue of star forming versus non-starforming — passive — galaxies. Kauffman *et al.* (2003) made a significant step by plotting the star formation rate (SFR), as measured by the broad-band index D4000 versus the galaxy mass, derived by fitting stellar population models to spectral energy distributions (SEDs) from SDSS multicolor photometry. Traditionally, galaxy color, used as a proxy for star formation rate, had been plotted against galaxy luminosity to show the well-known bimodality of red vs. blue galaxies first described by Baade (1958). Kauffman *et al.* showed something new, that galaxies in the contemporary universe are bimodal — star forming or passive — only if they have masses $\lesssim 10^{11}$ $M_{\odot}$. Above this mass, nearly all are passive.

The richly populated plot of D4000 vs stellar mass also showed a smaller population in between the passive "red-sequence" and the star forming "blue cloud." Beginning with Faber *et al.* (2005), dozens of studies identified these inhabitants of the "green valley" (GV) as "quenched galaxies" — those making their way from the blue cloud to the red sequence as the result of a generic event that truncated star formation.

Recently, Schwanski *et al.* (2014) have showed that ultraviolet colors of GV galaxies do not support the idea of a in-process transformation. Here we present other evidence

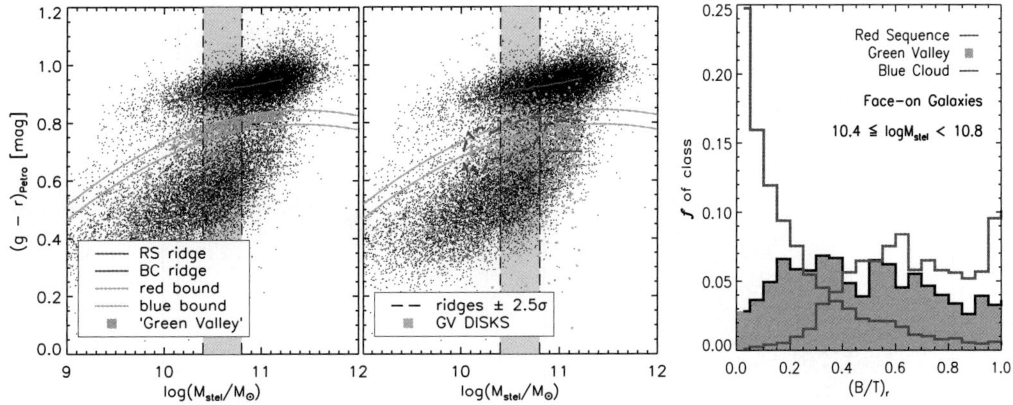

Figure 1. The left panel plots g-r color versus stellar mass, for the 'face-on' SDSS sample, showing the blue cloud and red sequence, and the 'Green Valley' (GV). The right panel shows the bulge-to-total light for these three distributions, limited to galaxies of mass 10.5 < log $M_{stellar}$ < 10.8, comparable to the Milky Way. Unlike the largely small-bulge galaxies of the blue cloud, most GV galaxies have substantial bulges, substantially explaining their intermediate color without a quenching process. The center panel repeats the left panel but with the bulges subtracted from the GV galaxies, showing that their starforming disks are those of normal starforming galaxies.

that this often-cited but unproven idea of 'galaxies in transition' is incorrect. Our hypothesis is that the primary reason galaxies appear in the GV is because they have normal star-forming disks but also substantial bulges which — viewed together — produce the intermediate color. This idea emerged from a study of the star forming main sequence (discussed below) by Abramson *et al.* (2014), which showed that taking the non-starforming bulges properly into account can resolve the puzzling feature of a "break" in this relation at higher masses — galaxies that are, in fact bulge dominated.

Returning to the GV, we also recognize that many of its inhabitants may simply be dusty, edge-on blue-cloud galaxies, so we remove them from our test sample by keeping only 'face-on' galaxies (ellipticity < 0.2). In Figure 1 (left) we plot (g - r) color as a function of stellar mass for an SDSS sample with z <0.20. We define the GV by the orange and blue lines, and, after excluding these points, define the red sequence and blue cloud — traced by the median lines. We take our measurements of bulge-to-total (r-band) light (B/T) from Simard *et al.* (2011). In Figure 1 (right), we plot the B/T distributions for galaxies of the red-sequence, the blue-cloud, and the GV for the mass limited sample indicated. As expected, the blue cloud galaxies are disk-dominated, the red sequence galaxies bulge-dominated, but the GV galaxies are intermediate, suggesting that, if they were "quenched" blue-cloud galaxies, strong bulge growth would have been required. We prefer the simpler explanation, that the vast majority of these GV galaxies are not in transition, but simply the combination of red bulges and blue disks. We demonstrate this in Figure 1 (center), in which the green points now show the GV galaxy colors after removing their bulges. — the disks of GV galaxies have normal blue-cloud colors.

The idea that *most* GV galaxies are "recently quenched and transiting" to the red sequence is shown to be incorrect by these data. However, it is reasonable to ask what fraction of the GV galaxies *could* have been recently quenched. Figure 2 (left) plots Lick-index Hδ absorption-line strengths versus star formation rate, from Hα measurements of SDSS fiber spectra, for the Figure 1 sample. There is a well- known relation of increasing Hδ absorption-line strength with SFR that is due to increasing fractions of hot, young stars. The scatter relates mostly to dust in these galaxies. The presence of strong Hδ

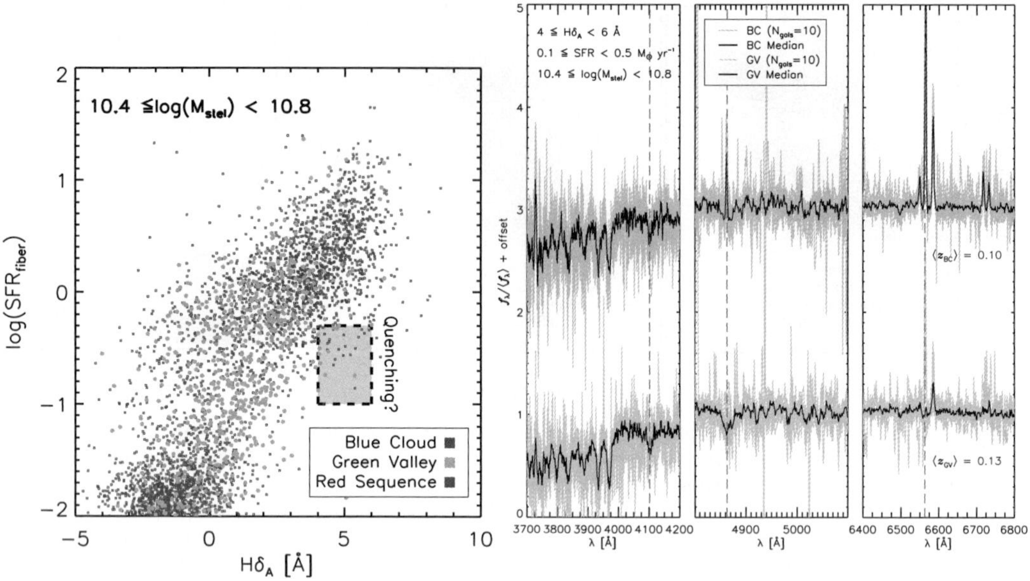

Figure 2. (left) The relationship of SFR and Hδ absorption for the $10.5 < \log \mathrm{M}_{stellar} < 10.8$ subsample of Figure 1 The grey box contains galaxies with strong-Hδ that are candidates for rapid quenching of star formation. (right) The composite spectra of blue cloud and GV galaxies in the grey box confirm that the latter matches the former except for a recent truncation of star formation, consistent with the idea that these galaxies are rapidly transitioning from the blue cloud to the red sequence. Quenched galaxies represent only a few percent of the GV galaxies in the SDSS sample.

in-excess of this relation is a signature of truncated star formation, or starbursts (e.g., Poggianti *et al.* 1999). In this diagram the GV galaxies generally track the blue-cloud galaxies, exhibiting the 1:10 ratio of these two populations. However, on the high-Hδ edge of this relation we see a proportion $\sim$1:1, meaning that there are in fact a larger fraction of Hδ-strong spectra. We select blue could and GV galaxies from a portion of this distribution, shown as the grey box in Figure 2 (left), and show the composite spectra of 10 GV and 10 blue-cloud galaxies in Figure 2 (right). Here, indeed, we see evidence of a quenched population. The color and absorption-line-strengths of the blue-could and GV composite are an excellent match, but the former has emission lines — most notably, Hα and Hβ — that are weak or absent in latter. This is exactly what is expected for quenched systems, where the 1-10 Myr-old stars that power the H II regions have disappeared, but the stellar population $\tau > 100$ Myr remains unchanged. Over a timescale $\tau \sim$1 Gyr, these GV galaxies should join the red sequence.

We estimate that, for the SDSS sample, the size of this quenched population is a few percent of the starforming population. In her talk at this Symposium, Professor Sandra Faber referred to these cases as "fast-track" quenching. She acknowledged that most galaxies in the GV are following the "slow-track," those with a time scale of several Gyr — indistinguishable from normal galaxy evolution, and not the result of a "quenching event." We conclude that for present-epoch galaxies the GV is basically composed of intermediate-bulge spiral galaxies that are slowly, steadily making their way from the blue cloud to the red sequence. through the normal process of gas exhaustion. Vulcani *et al.* (2014) come to similar conclusions in their study comparing the colors and morphologies of PM2GC galaxies ($z \sim 0.1$) in the GV, red sequence, and blue cloud. counterparts.

Professor Faber also suggested that "fast-track" quenching might be a larger component of early universe galaxies. We note that the situation at intermediate-redshift is similar to what we have discussed here, with an additional ten-times larges population of starburst and poststarburst galaxies (e.g., Dressler *et al.* 2013). One might think that these are cases in which star formation has been stopped — by means of a starburst, but in Dressler *et al.* and Abramson *et al.* (2013) we provide strong evidence that most starbursts do not transform blue cloud galaxies to red. If they did, the number of intermediate-z galaxies (log $M_{stellar} > 10.2$) on the red sequence would be growing much faster than observed. Rather, we propose that these starbursts are the result of minor mergers, and that when they occur in star forming galaxies, they return to same. Likewise, when a red-sequence galaxy suffers a minor merger, it generally returns to its passive state. We think that only a major merger is likely to lead to a starbust-mediated transformation from a starforming to passive galaxy.

3. A Better Parameterization of Star Formation Histories?

The IMACS Cluster Building Survey team, including the authors of this paper plus Gus Oemler, Mike Gladders, Bianca Poggianti, and Benedetta Vulcani, is pursuing the question of the possible forms of the SFHs of galaxies. A common approach to this issue is to consider the 'Star Forming Main Sequence' (SFMS: Noeski *et al.* 2007; Whitaker *et al.* 2012), a plot of an evolving relation of SFR and stellar mass, as a template for galaxy evolution (e.g. Peng *et al.* 2010). If the scatter were small — possibly consistent with observational errors, as suggested in these studies — it might be possible to reduce SFHs to a single path, one that connects the SFR uniquely to a mass at any given epoch. On the other hand, if the scatter is real, then "connecting the dots" from the SFMS of one epoch to another is a difficult if not hopeless task.

The ICBS team has assembled evidence that the scatter in the SFMS *is* real, and large — approaching 0.5 dex if the sample is not "pruned" and when selection effects are properly taken into account (Oemler *et al.* 2014; see also Salim *et al.* 2007). We believe this is strong evidence for diverse SFHs for galaxies at any mass. Oemler *et al.* (2013) compared the distribution of specific star formation rates, sSFRs, of ICBS field galaxies to present-epoch field galaxies and found that, at $z = 0.4$, 20% of all galaxies could not be fit by traditional τ-models, that is, they exhibit *rising* star formation rates at that epoch. Searching for an analytical form that can fully fit such diversity in SFHs, we noted, as described in Gladders *et al.* (2013), that the Madau plot — the SFH of the universe itself — is well fit by a lognormal. We show this here using the new compilation of star formation rate densities from Madau and Dickinson (2014) in Figure 3 (left), where the authors have included a polynomial fit, which we imagine as physically unmotivated. An arguably better fit is achieved by the lognormal, a simple form that occurs frequently in natural systems and is described by only two, easily understandable parameters, in this case t_0 — the time of peak star formation, and τ — the half-width of the SFH.

If the lognormal is a good fit to the universal SFH, might it be a good description of individual SFHs as well? Gladders *et al.* (2013) tested that supposition by using as constraints the sSFR distribution composed of two $z \approx 0$ samples and the Madau plot itself, and requiring each of the present-epoch galaxies to have a SFH that is lognormal in time. The model, constrained only by the local sample and the Madau relation, correctly predicts the distribution of sSFR for the ICBS and AEGIS samples at $z = 0.2$, 0.4, 0.6, and 0.8. (Gladders *et al.* — see Figure 7), in contrast to an attempt using Gaussian SFHs that fails badly (*ibid*, Figure 12). Adding in the sSFHs for the intermediate-redshift data produces modestly better constraints: Figure 3 (right) shows the distribution of (t_0, τ)

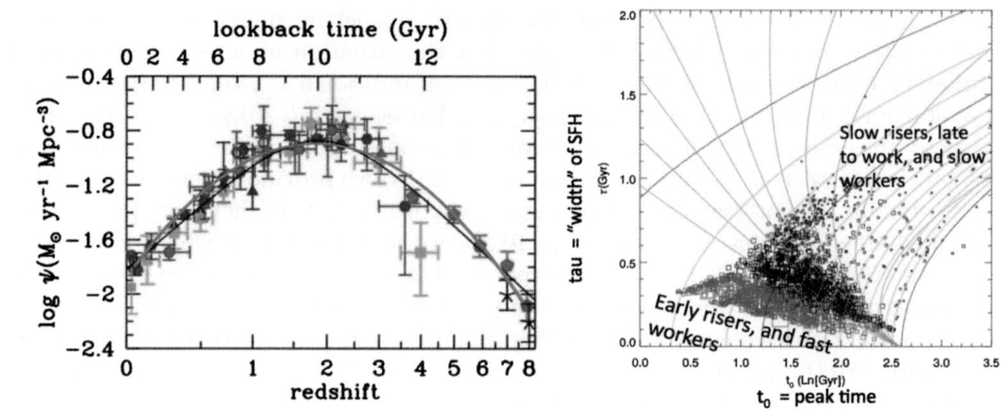

Figure 3. (left) The "Madau diagram," from Madau and Dickinson (2014). The thick blue line is a lognormal fit with $t_0 = 4.66$ Gyr and $\tau = 1.78$ Gyr, which is arguably a better fit to the data than the polynomial constructed by those authors. (right) The distribution of (t_0, τ) parameters for the lognormals describing the evolution of the present-epoch field sample. Passive galaxies, defined by Gladders *et al.* as SFR < 0.05 M$_\odot$ yr^{-1} form the sequence of red points, while the starforming galaxies scatter, exhibiting a wide range of SFHs.

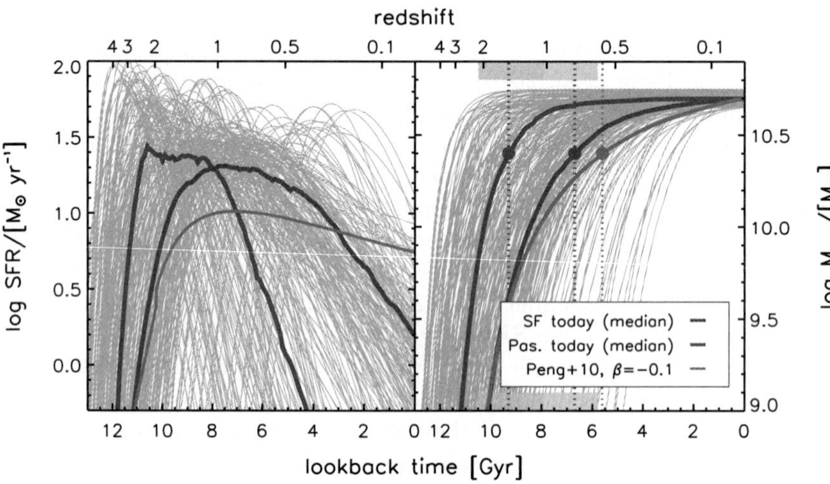

Figure 4. (left) Lognormal SFHs (left) and mass-growth curves (right) for today's Milky Way–mass galaxies, from the model by Gladders *et al.* (2013). Thin orange/blue lines, respectively, denote currently "passive" ($\log \mathrm{sSFR/yr^{-1}} < 11$) or "starforming" (sSFR above that threshold) galaxies. Thick red/blue tracks show median SFR(t) for each class. While, on average, these populations had very different SFHs, our modeling predicts a wide diversity of growth histories *within* each class (see vertical lines/banding at right denoting their median/inter-quartile spread in half-mass times). Such diversity is not a feature of models such as that of Peng *et al.* (2010) — which yield a single track for all starforming objects at fixed stellar mass, today (pink curves) — and should be testable with advanced methods of stellar-population synthesis.

for this final fit of Gladders *et al.* As we will show in an upcoming paper (Abramson *et al.* 2014), using only these $z < 1$ data we can fit the state-of-the-art mass functions of Tomczak *et al.* (2014) of both passive and star forming galaxies for $0.5 < z < 2.5$, even though mass was never used to inform the SFH modeling.

In summary, a model including a wide diversity of lognormal SFHs reproduces today's three common relations that record the evolution of basic galaxy parameters over cosmic time: the Madau plot of SFR-density; the SFMS *with scatter*; and the mass functions of both starforming and passive galaxies. However, our success is not unique, as Peng *et al.*, Kelson (2014), and Behroozi and collaborators (Behroozi *et al.* 2013a,b) have all claimed the same for very different descriptions of SFHs (see also Simha *et al.* 2014). The galaxies in these studies are represented as numbers — measures of basic or even integral properties, for points that cannot in general be connected over time. It is not surprising, then, that the Madau diagram, SFMS, and mass functions are insufficient to choose between different models of SFHs. Our formalism allows for much more diversity than simply "starforming" *or* "quenched." Figure 4 (left) is a plot showing the diversity of our lognormal SFHs, both as aggregate SFHs for star forming galaxies contrasted to today's passive galaxies, and for individual SFHs. Also plotted is the single SFH of a galaxy that is star forming today in the Peng *et al.* model. Figure 4 (right) shows the mass growth derived for these SFHs. Even in the average of lognormal histories, their is substantial difference in galaxy growth over time — the SFHs for galaxies at the same mass can be very diverse and still fit available data.

The challenge is clear: we need to move beyond these three basic parameterizations and test the SFHs of individual galaxies, as we have begun to do with the Milky Way and its neighbors. Only when such detailed measurements replace the general, population-based data we have today will the full picture of galaxy star formation histories emerge.

Acknowledgements

The authors thank the ICBS team for many intense and inspiring discussions.

References

Abramson, L. E., Dressler, A., Gladders, M. D., *et al.* 2013, *ApJ*, 777, 124
Abramson, L. E., Gladders, M. G., & Dressler, A., *et al.* 2014, in preparation.
Baade, W. 1958, *Ricerche Astronomiche*, 5, 3
Behroozi, P. S., Wechsler, R. H., & Conroy, C. 2013a, *ApJ*, 770, 57
Behroozi, P. S., Marchesini, D., Wechsler, R. H., *et al.* 2013b, *ApJ*, 777, L10
Dressler, A., Oemler, A., Jr., Poggianti, B. M., *et al.* 2013, *ApJ*, 770, 62
Gladders, M. D., Oemler, A., Dressler, A., *et al.* 2013, *ApJ*, 770, 64
Kelson, D. D. 2014, arXiv:1406.5191
Noeske, K. G., Weiner, B. J., Faber, S. M., *et al.* 2007, *ApJ*, 660, L43
Oemler, A., Jr., Dressler, A., Gladders, M. G., *et al.* 2013, *ApJ*, 770, 63
Oemler, A., Jr., Abramson, L. E., Gladders, M. G., *et al.* 2014, in preparation.
Madau, P. & Dickinson, M. 2014, *ARA&A*, 52, 415
Peng, Y.-j., Lilly, S. J., Kovač, K., *et al.* 2010, *ApJ*, 721, 193
Poggianti, B. M., Smail, I., Dressler, A., *et al.* 1999, *ApJ*, 518, 576
Simard, L., Mendel, J. T., Patton, D. R., Ellison, S. L., & McConnachie, A. W. 2011, *ApJS*, 196, 11
Schawinski, K., Urry, C. M., Simmons, B. D., *et al.* 2014, *MNRAS*, 440, 889
Simha, V., Weinberg, D. H., Conroy, C., *et al.* 2014, arXiv:1404.0402
Tomczak, A. R., Quadri, R. F., Tran, K.-V. H., *et al.* 2014, *ApJ*, 783, 85
Vulcani, B., Poggianti, B. M., Fritz, J., *et al.* 2014, submitted to ApJ. Abramson, L. E., Gladders, M. G., *et al.* 2014, in preparation.
Whitaker, K. E., van Dokkum, P. G., Brammer, G., & Franx, M. 2012, *ApJ*, 754, L29

Galaxy Masses as Constraints of Formation Models
Proceedings IAU Symposium No. 311, 2014
M. Cappellari & S. Courteau, eds.

doi:10.1017/S1743921315003567

Kinematic Evolution of Field and Cluster Spiral Galaxies

Bodo L. Ziegler[1] **and Asmus Böhm**[2]

[1]University of Vienna, Department of Astrophysics, Türkenschanzstr. 17, 1180 Vienna, Austria
email: bodo.ziegler@univie.ac.at

[2]Institute for Astro- and Particle Physics, Technikerstrasse 25/8, 6020 Innsbruck, Austria
email: asmus.boehm@uibk.ac.at

Abstract. We investigate the evolution of the Tully–Fisher relation out to $z = 1$ with 137 emission-line galaxies in the field that display a regular rotation curve. They follow a linear trend with lookback time being on average brighter by 1.1 Bmag and 60% smaller at $z = 1$. For a subsample of 48 objects with very regular gas kinematics and stellar structure we derive a TF scatter of 1.15mag, which is two times larger than local samples exhibit. This is probably due to modest variations in their star formation history and chemical enrichment.

In another study of 96 members of Abell 901/902 at $z = 0.17$ and 86 field galaxies with similar redshifts we find a difference in the TFR of 0.42mag in the B-band but no significant difference in stellar mass. Comparing specifically red spirals with blue ones in the cluster, the former are fainter on average by 0.35 Bmag and have 15% lower stellar masses. This is probably due to star formation quenching caused by ram-pressure in the cluster environment. Evidence for this scenario comes from strong distortions of the gas disk of red spirals that have at the same time a very regular stellar disk structure.

Keywords. galaxies: spiral - galaxies: evolution

1. Introduction

Internal kinematics of galaxies provide important clues about their nature and reveal physically distinct components (e.g. Kalinova *et al.* 2014). While stars are on colissionless orbits tracing the gravitational potential dominated by dark matter, gas clouds can react faster on subtle effects and may display peculiar motions revealing interaction events. In case of undisturbed spiral galaxies rotation produces a characteristic velocity profile with a steep rise in the inner part turning over to a flat part, the so-called rotation curve RC. In such a virialized state, the maximum velocity $V_{\rm max}$ provides a good estimate of the total mass of a galaxy and tightly correlates with the luminosity in the Tully–Fisher relation TFR. Under this assumption the TFR is an important scaling relation to probe quantitatively the evolution of disk galaxies with redshift as applied in several studies before (e.g. Böhm & Ziegler 2007; Miller *et al.* 2011). Since stellar kinematics from absorption lines requires a higher S/N most studies of distant galaxies rely on emission line measurements from the warm ionized gas. Here, we report on two applications: a TFR study of distant field galaxies and an investigation of transition objects in the cluster complex Abell 901/2.

2. Evolution of the Tully–Fisher relation to z = 1

Continuing our previous investigations we expanded our sample towards fainter galaxies with new observations with ESO's VLT using FORS (Böhm & Ziegler 2014). The whole sample holds high-quality spectra of 238 objects. For 137 disk galaxies we were

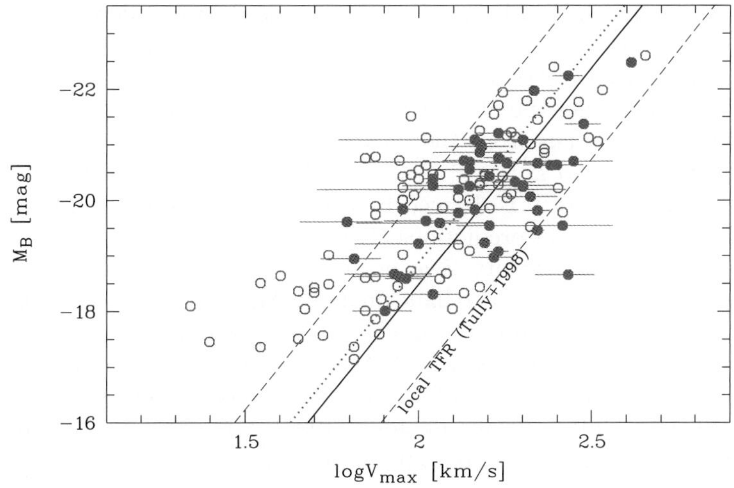

Figure 1. The Tully–Fisher relation of 137 field galaxies with $0.05 < z < 0.97$ (open symbols) compared to the local relation from Tully *et al.* 1998 (solid line & 1-σ boundaries dashed). Filled circles represent a subsample of 48 disk galaxies with low morphological asymmetry and high quality rotation curves. Fitting their zeropoint yields the dotted line offset by 0.40mag towards brighter luminosities.

able to extract RCs and determine $V_{\rm max}$, while 101 emission-line objects were rejected from our TF analysis due to kinematic distortions. These field spirals are spread like $0.05 < z < 0.97$ with a median redshift of 0.45 corresponding to lookback times between 0.6 and 7.6 Gyr and $\langle t_{\rm lb} \rangle = 4.5$ Gyr. In Fig. 1 we show their TFR in rest-frame B-band compared to the local study of Tully *et al.* (1998). The slope of the distant sample is with $-7.16^{+0.71}_{-0.53}$ similar to the local value of -7.79. We further create a subsample of 48 objects restricted to have both high-quality RCs and a low structural asymmetry ($A_{\rm morph} < 0.25$ in the CAS system of Conselice at al. (2000)) to ensure regularity in both the gas and the stellar component in order to avoid any non-relaxed contribution to the

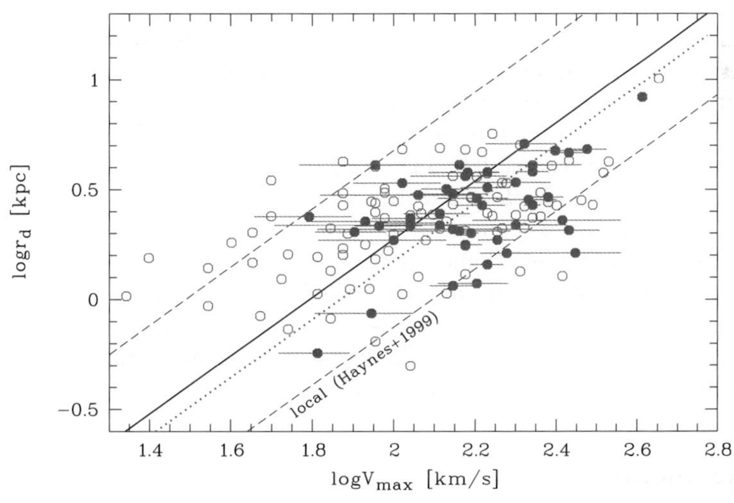

Figure 2. The Velocity–Size relation of the same 137 galaxies like in Fig. 1 this time compared to the local relation from Haynes *et al.* (1999). On average, distant galaxies have disk scale lengths r_d smaller by 0.15 dex than local ones for given $V_{\rm max}$ (fit indicated by dotted line).

velocity field. Assuming the local slope this subsample shows moderate brightening of on average $\langle \Delta M_B \rangle (z \approx 0.5) = -0.40^m$ and a scatter of 1.15^m, which is two times larger than the local one. This may be attributed to a modest diversity of star formation histories and chemical enrichments of the observed galaxies (Ferreras *et al.* 2014). Looking at the evolution of the TF residuals from the local relation we fit our whole sample linearly with lookback time and find $\Delta M_B = -(3.63 \pm 1.78) \log(1+z)$ albeit with large scatter. Such an increase in luminosity is derived by other studies, too (e.g. Miller *et al.* 2011), and in accordance with model predictions (Dutton *et al.* 2011).

We also investigate the Velocity–Size evolution of the same galaxies by measuring their disk scale lengths r_d in HST I-band images. Although the scatter is rather large, our sample is offset by -0.15 dex to smaller r_d for given $V_{\rm max}$ compared to the local sample of Haynes *et al.* (1999), see Fig. 2. Note that the local sample was observed in I-band, too, and the scale lengths of the distant galaxies were slightly corrected for rest-frame wavelength shift following de Jong (1996). On average, disk sizes decrease by $\Delta \log r_d = -(0.59 \pm 0.38) \log(1+z)$ becoming about 60% smaller at $z = 1$, which is again in accordance with models.

3. Red spirals as stripped transition objects in clusters

Looking for environmental effects we examine galaxies in the multiple cluster system Abell 901/902 at $z = 0.17$ (Gray *et al.* 2009). An observational program with ESO's VLT using VIMOS targeted ~ 200 emission-line galaxies in the cluster complex with tilted slits placed along the major axis (PI A. Böhm, see Bösch *et al.* (2013a)). A kinematic analysis was possible for 96 cluster members and 86 field spirals with a similar average redshift $(0.12 < z < 0.38)$. However, many objects don't exhibit a regular RC but distorted kinematics. To quantify the degree of distortion we introduce similar to Dale *et al.* (2001) an RC asymmetry index $A_{\rm RC}$ based on differences of the approaching and receding arms. We find an average value of $\langle A_{\rm RC} \rangle = 17\%$ in the field, while distortions are more frequent in the cluster, yielding $\langle A_{\rm RC} \rangle = 25\%$. Restricting to objects with high-quality RCs for a TF analysis we find that cluster members are on average modestly fainter in rest-frame B-band by $\Delta M_B = 0.42 \pm 0.15^m$. However, there is no significant difference in the stellar TF with $\Delta \log M_* = 0.00 \pm 0.07$.

Further, we concentrate on red spirals that were defined by Wolf *et al.* (2009) to have red colors similar to red-sequence galaxies but (spiral) disk morphology and significant star formation rates (SFRs), albeit lower than typical for blue spirals. Those red spirals with regular RCs are on average fainter in the B-band TFR by $\Delta M_B = 0.35 \pm 0.12^m$ and show a slight difference in the stellar TF by $\Delta \log M_* = 0.06 \pm 0.05$ compared to blue-cloud cluster members (Bösch *et al.* 2013b). These results may be explained by a recent quenching of SF for red spirals caused by ram-pressure stripping in the cluster removing gas. This interpretation is strengthened by Fig. 3: many red spirals show strong rotation curve asymmetries combined with an undisturbed stellar morphology. The only cluster-specific interaction process which affects the gas disk but not the stellar morphology is ram-pressure (Kronberger *et al.* 2008). Thus, our analysis adds evidence that red spirals are transition objects between field blue-cloud and cluster lenticular galaxies.

Acknowledgements

Based on ESO PID 081.B-0107, 384.A-0813 and other programs. AB acknowledges support by the Austrian Science Foundation FWF (grants P19300-N16 and P23946-N16). BZ thanks SOC and LOC for a very nice symposium.

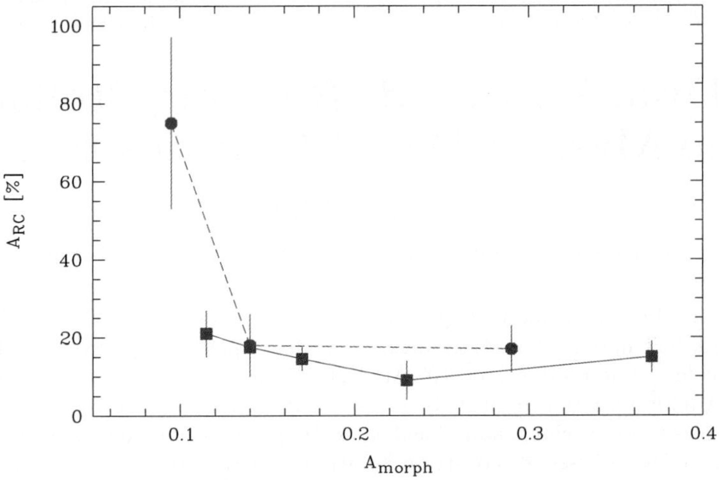

Figure 3. Rotation curve asymmetry $A_{\rm RC}$ versus morphological asymmetry $A_{\rm morph}$ for red spirals (circles) and blue spirals (squares) in the cluster system A901/902. Average error-weighted values have been computed for bins holding 9 (red) and 12 objects (blue), respectively. A large fraction of red spirals show strong rotation curve asymmetries, i.e. disturbed gas kinematics, combined with undistorted stellar structure (i.e. low $A_{\rm morph}$). This can best be explained by ram pressure due to the intra-cluster medium.

References

Böhm, A. & Ziegler, B., 2007, *ApJ*, 668, 846.
Böhm, A. & Ziegler, B., 2014, *A&A*, in prep.
Bösch, B., Böhm, A., ..., Ziegler, B., *et al.*, 2013a, A&A 549, A142
Bösch, B., Böhm, A., ..., Ziegler, B., *et al.*, 2013b, A&A 554, A97
Conselice, *et al.*, 2000, *ApJ*, 529, 886
Dale, *et al.*, 2001, *AJ*, 121, 1886
de Jong, R. S. 1996, å, 313, 377
Dutton, *et al.*, 2011, *MNRAS*, 410, 1660
Ferreras, *et al.*, 2014, *MNRAS*, 437, 1872.
Gray, *et al.*, 2009, *MNRAS*, 393, 1275.
Haynes, *et al.*, 1999, *AJ*, 117, 1668
Kalinova, *et al.*, 2014, *A&A*, in prep.
Kronberger, *et al.*, 2009, å, 483, 783
Miller, *et al.*, 2011, *ApJ*, 741, 115
Tully, *et al.*, 1998, *AJ*, 115, 2264
Wolf, *et al.*, 2009, *MNRAS*, 393, 1302

Galaxy Masses as Constraints of Formation Models
Proceedings IAU Symposium No. 311, 2014
M. Cappellari & S. Courteau, eds.

doi:10.1017/S1743921315003579

The Dwarfs Beyond: Relating Stellar and Halo Mass in Dwarf Galaxies to $z \sim 1$

Sarah H. Miller
University of California, Irvine, email: shmiller@uci.edu

Abstract. Do the kinematics and mass profiles of dwarf galaxies present a fundamental challenge to standard cold dark matter (CDM) models? New, deep spectroscopy using DEIMOS on Keck for hundreds of low stellar mass (10^7-10^9 M$_\odot$) star-forming galaxies at intermediate redshift ($0.2 < z < 1$) addresses this inquiry in a way that is less subject to cosmic variance and environmental bias than previous, more local work. Half of this sample reveals resolved, doppler-shifted nebular emission, used to constrain rotation curves. From these we can construct the stellar mass Tully-Fisher relation to masses as low as $\sim 10^7$ M$_\odot$, and a persistent discrepancy is found between predictions from simulations and models compared to our observations. We suggest on-going and future tests that will be more effective in distinguishing between the effects of baryonic feedback and alternative models of dark matter in this remarkable regime.

Keywords. galaxies: dwarfs - galaxies: evolution - galaxies: formation

1. Introduction

The foremost challenge to the favored dark matter paradigm (WIMP-like CDM) is the discrepancy between the predicted inner densities of galaxies compared to those observed (Boylan-Kolchin *et al.* 2011; Ferrero *et al.* 2012; Weinberg *et al.* 2013; Papastergis *et al.* 2014). Revealed across the shapes and amplitudes of rotation curves, this discrepancy is particularly clear in low-mass galaxies where dark matter is predicted to dominate even their inner-most potential wells – much more so than in massive galaxies. Before Miller *et al.* (2014, M14) directly resolved dwarf rotation curves to intermediate redshifts ($0.2 < z < 1$), this issue had only been explored in the low-redshift, local universe. The limits of earlier samples subjected results to cosmic variance issues, and the local range of environments left open the possibility that dwarf galaxy densities were discrepant with theoretical predictions because of their complicated interaction histories in our particular cosmic neighborhood (or that this was simply a case of "small-number statistics").

To surmount cosmic variance, limited statistics, and environmental caveats, we have been leading a novel study of directly-resolved internal kinematics of some of the lowest mass galaxies at the highest redshifts ever targeted. We are amassing a large, rich dataset selecting from Hubble Space Telescope imaging and acquiring deep Keck DEIMOS spectroscopy that will serve as a valuable archive for years to come, not only for kinematics, but also for exploring the exotic interstellar medium environments of these dwarfs, including both emission and absorption feature diagnostics. Typically we find that half of the observed sample, selected simply by photometric redshift ($0.2 < z < 1$), stellar mass ($\log M_*/M_\odot < 9.3$), and magnitude ($i_{AB} > 25$), presents significant signal-to-noise for both spatially and spectrally resolving emission with DEIMOS after 6-10 hours of observing on Keck in good conditions. With hundreds of new dwarfs now observed, we have found persistent inconsistencies between abundance/simulation methods and the observational kinematic method (using the extrapolated weak lensing relation of Reyes *et al.* 2012, to convert to halo mass) that cannot be explained by the classic understanding

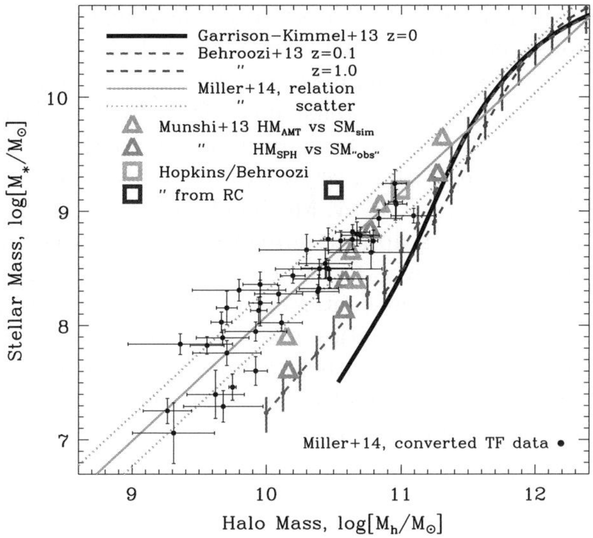

Figure 1. The stellar-to-halo mass relation according to the latest abundance matching curves (navy, aqua and red curves at z=0.0, 0.1 and z=1.0, of Garrison-Kimmel *et al.* (2013) & Behroozi *et al.* (2013), respectively) derived from CDM simulations matched to observed stellar mass functions, compared to that inferred for our sample of intermediate redshift dwarf galaxies by converting the measured rotation velocities in the Tully-Fisher relation to halo masses using the procedure discussed in M14 (summarized in the text). Black points represent the data, the solid grey line is the best-fit relation and the scatter is indicated with dotted grey lines. The triangles and squares denote various numerical simulations (Munshi *et al.* 2013, P. Behroozi & P. Hopkins - private communication).

of the missing satellites or the cusp-core problems (see Fig. 1 for the pilot sub-sample in M14). With this sample representing star-forming galaxies in group, field, and void environments from over half the age of the universe, these early results help us more precisely define questions to identify what is missing from our current theoretical and observational picture of both galaxy evolution and dark matter.

2. Extending the Measurement of the M_* Tully-Fisher Relation

A tripling of the M14 pilot sample will be published later this year (Miller *et al.*, in prep), offering a more robust assessment of the scatter of the stellar mass Tully-Fisher relation to lower mass, which stands as one of the more limiting uncertainties in connecting observable galaxies to their dark halos. A wide range of techniques has been used in the past decade or two to make this connection, from halo occupation distribution modeling (Berlind & Weinberg 2002; Wechsler *et al.* 2006) to more simplified abundance matching techniques (Behroozi *et al.* 2013). While a consistent picture can be constructed over the mass ranges typically explored in extra-galactic surveys to intermediate redshifts (where log $M_*/M_\odot > 9$), this picture begins to diverge at lower masses. A favored explanation is the increased scatter of stellar and gas fractions in dwarf-regime halos, however this has yet to be well constrained observationally.

The problem with comparing observations to models using axes such as those in Fig. 1 is that halo mass is a particularly indirect measurement and the assumptions made to place curves and symbols on such a plot may not ultimately be consistent with each other (or with nature, for that matter). For instance in the case of the converted TF data points, optically measured velocities have been converted to halo velocities using an extrapolated weak lensing relation, which are then converted to a halo mass. An important next step will be to conduct consistency tests between models and observations on more directly observable measurements, like enclosed mass at radii which are directly observable, such as various optical radii, or simply the stellar mass Tully-Fisher relation itself.

Why does the Tully-Fisher relation exist in its observed shape and scatter? Neither vanilla ΛCDM, nor calculations with naive gravitational physics give satisfactory answers (Courteau *et al.* 2007; Dutton *et al.* 2011; McGaugh 2012), and narrow implementations of modified Newtonian dynamics appear unrealistic by independent lines of evidence. After 35 years of seeking this answer, are we on the brink of understanding what drives the evolution of not just the slope and intercept of this unique scaling relation, but very importantly, of the scatter? Is this relation a key piece of the dark matter puzzle, or are we simply seeing how baryons scale with baryonic effects where they matter most, in the inner regions of galaxies, even down to masses far lower than implementations of the CDM paradigm have typically predicted?

3. Future Opportunities

Looking long-term, the ultimate galaxy laboratory for distinguishing between baryonic feedback effects and fundamentally-altered dark matter physics from the vanilla CDM implementation will be the unique performance regime of the next generation of giant segmented mirror telescopes and the James Webb Space Telescope. These telescopes will push well into the sensitivity and resolution regime needed to trace the dynamics of many galaxies, both isolated and interacting, of lower stellar mass and slower rotation rates at higher redshifts than what has previously been possible on Earth or in space. By tracking the evolving shapes of supposedly dark matter dominated galaxy rotation curves through the age of the universe, we will be tracking the transfer of star-formation energy to the mass redistribution. Both the low-mass and high-redshift regimes of galaxy formation, with their limited scope of star-formation, reveal particularly strong discriminators between feedback effects and intrinsic dark matter properties.

Complete dynamical characterization of this regime (dwarf galaxies to high-redshift) must be conducted while accounting for environment, making up the three prongs of an ideal future survey:

(*a*) **Low-Mass Regime** ($\log M_*/M_\odot < 8$, especially $\log M_*/M_\odot < 6$) to be squarely in the dark matter -dominated regime where baryons and star-formation have a limited share of the energy budget, selected as simply as possible and ideally where completeness is well understood.

(*b*) **Evolution Sensitivity** (to $z \sim 1$ and beyond) to track the smoothness of density profile change across both time and environment, as well as star-formation.

(*c*) **Environmental Variation** (from the densest environments to voids) to compare those galaxies which would have had limited interaction to those that are entirely governed by their interaction history, to better assess the driving forces in the internal dynamics of dwarf galaxies.

Will we see the statistical spread of erratic baryonic feedback redistributing cold dark matter particles in episodic jumps according to environment? Or might we see the telltale profile evolution of warm dark matter, or even perhaps the progressive coring-out of profiles from self-interacting dark matter? In the latter possibility, the rate of profile evolution may even constrain the interaction cross-section of dark matter. Without a wide-range sampling in mass, time, and environment for spatially and spectrally resolved kinematics, answers will remain elusive.

Just as important as the type of observational campaign proposed is the continued development of well-understood, well-controlled models of both standard and alternative physics of both the baryonic and dark matter components of galaxies. If we have learned anything in the past decade of galaxy evolutions studies, it is that the relationship between baryonic and dark matter in galaxies is non-trivial. Hydrodynamic, analytic, and semi-analytic models all contribute to the framework we use to interpret our observations. For the types of comparison studies needed in this work, which are some of the most direct and detailed comparisons we can make in the dark-baryon connection, the focus for model building must be the physical processes that affect the *observable* part of the rotation curve. A huge challenge in the next generation of model-building is to resolve the observables predicted from physically relevant processes. While difficult, some studies have begun to show this is possible. The continued, creative development of resolved, and *observably relevant* models, in preparation for the exciting observational regimes on the horizon, will be a worthwhile investment.

Acknowledgements

SHM acknowledges support from the University of California - Irvine Chancellor and ADVANCE program, the McCue Fellowship of the UCI Physics & Astronomy Department, and the California Institute of Technology. Further thanks to the University of Oxford in supplementing costs associated with attending IAU Symposium No. 311.

References

Behroozi, P. S., Wechsler, R. H., & Conroy, C. 2013, *ApJ*, 770, 57
Berlind, A. A. & Weinberg, D. H. 2002, *ApJ*, 575, 587
Boylan-Kolchin, M., Bullock, J. S., & Kaplinghat, M. 2011, *MNRAS*, 415, L40
Courteau, S., Dutton, A. A., van den Bosch, F. C., MacArthur, L. A., Dekel, A., McIntosh, D. H., & Dale, D. A. 2007, *ApJ*, 671, 203
Dutton, A. A., *et al.* 2011, *MNRAS*, 410, 1660
Ferrero, I., Abadi, M. G., Navarro, J. F., Sales, L. V., & Gurovich, S. 2012, *MNRAS*, 425, 2817
Garrison-Kimmel, S., Rocha, M., Boylan-Kolchin, M., Bullock, J. S., & Lally, J. 2013, *MNRAS*, 433, 3539
McGaugh, S. S. 2012, *AJ*, 143, 40
Miller, S. H., Ellis, R. S., Newman, A. B., & Benson, A. 2014, *ApJ*, 782, 115
Munshi, F., *et al.* 2013, *ApJ*, 766, 56
Papastergis, E., Giovanelli, R., Haynes, M. P., & Shankar, F. 2014, (arXiv:1407.4665)
Reyes, R., Mandelbaum, R., Gunn, J. E., Nakajima, R., Seljak, U., & Hirata, C. M. 2012, *MNRAS*, 425, 2610
Wechsler, R., Zentner, A., Bullock, J. S., Kravtsov, A., & Allgood, B. 2006, *ApJ*, 652, 71
Weinberg, D. H., Bullock, J. S., Governato, F., Kuzio de Naray, R., & Peter, A. H. G. 2013, (arXiv:1306.0913)

Author Index